MODERN INTERDISCIPLINARY UNIVERSITY STATISTICS EDUCATION

Proceedings of a Symposium

Committee on Applied and Theoretical Statistics
Board on Mathematical Sciences
Commission on Physical Sciences, Mathematics, and Applications
National Research Council

National Academy Press
Washington, D.C. 1994

Support for this project was provided by the National Science Foundation under Grant No. DMS-9221287. Any opinions, findings, conclusions, or recommendations expressed in this material are those of the authors and do not necessarily reflect the views of the National Science Foundation.

Library of Congress Catalog Card Number 94-65037
International Standard Book Number 0-309-05033-2

Additional copies of this report are available from:
National Academy Press, Box 285
2101 Constitution Avenue, N.W.
Washington, D.C. 20418
800-624-6242
202-334-3313 (in the Washington metropolitan area)

B-286

Printed in the United States of America

COMMITTEE ON APPLIED AND THEORETICAL STATISTICS

JON R. KETTENRING, Bell Communications Research, *Chair*
YVONNE BISHOP, U.S. Department of Energy
MARY ELLEN BOCK, Purdue University
CLIFFORD C. CLOGG, Pennsylvania State University
MARJORIE G. HAHN, Tufts University
DOUGLAS M. HAWKINS, University of Minnesota
DAVID G. HOEL, Medical University of South Carolina
KARL E. PEACE, Biopharmaceutical Research Consultants
STEPHEN M. POLLOCK, University of Michigan
DARYL PREGIBON, AT&T Bell Laboratories

Staff
JOHN R. TUCKER, Senior Program Officer

BOARD ON MATHEMATICAL SCIENCES

COMMISSION ON PHYSICAL SCIENCES, MATHEMATICS, AND APPLICATIONS

Competent statisticians will be front line troops in our war for survival — but how do we get them? I think there is now a wide readiness to agree that what we want are neither mere theorem provers nor mere users of a cookbook. A proper balance of theory and practice is needed and, more important, statisticians must learn to be good scientists, a talent which has to be acquired by experience and example.

— George E. P. Box,
"Science and Statistics"

Preface

At its August 1992 meeting in Boston, the Committee on Applied and Theoretical Statistics (CATS) noted widespread sentiment in the statistical community that upper-level undergraduate and graduate curricula for statistics majors and postdoctoral training for statisticians are currently structured in ways that do not provide sufficient exposure to modern statistical analysis, computational and graphical tools, communication skills, and the ever-growing interdisciplinary uses of statistics. Approaches and materials once considered standard are being rethought. The growth that statistics has undergone is often not reflected in the education that future statisticians receive. There is a need to incorporate more meaningfully into the curriculum the computational and graphical tools that are today so important to many professional statisticians. There is a need for improved training of statistics students in written and oral communication skills, which are crucial for effective interaction with scientists and policy makers. More realistic experience is needed in various application areas for which statistics is now a key to further progress.

In response to this sentiment, CATS initiated a project on modern interdisciplinary university statistics education. With support from the National Science Foundation, CATS organized and held a one-and-one-half-day symposium on that topic in conjunction with the August 1993 San Francisco Joint Statistical Meetings. The symposium's focus was what changes in statistics education are needed to (1) incorporate interdisciplinary training into the upper-undergraduate, graduate, and postdoctoral statistics programs, (2) bring the upper-undergraduate and graduate statistics curricula up to date, and (3) improve apprenticing of statistics graduate and postdoctoral students and appropriately reward faculty mentors.

These proceedings have been compiled to capture the timely and important presentations and discussions that took place at that symposium. It should be noted that the opinions expressed in this volume are those of the speakers or discussants and do not necessarily represent the views of CATS or of the National Research Council. It is hoped that these presentations and discussions will not only initiate a process of long overdue change in upper-undergraduate, graduate, and postdoctoral education for statisticians, but will also stimulate the incorporation of interdisciplinary experience and realistic apprenticing in the nation's programs for statistical science majors, advanced degree candidates, and postdoctoral students. It is also hoped that this changing of the nation's statistics education programs will benefit the nation in a larger sense by serving as a model for other disciplines, such as mathematics and the sciences, to emulate.

Contents

Opening Remarks 1
John R. Tucker, National Research Council

PART I

What the Customer Needs in Statisticians
Session Chair: Jerome Sacks, National Institute of Statistical Sciences

What Industry Needs 5
Jon R. Kettenring, Bell Communications Research

What Academia Needs 10
Peter J. Bickel, University of California at Berkeley

What the Government Needs 14
N. Phillip Ross, Environmental Protection Agency

A Larger Perspective 18
John C. Bailar III, McGill University

Discussion 21

The Content of Courses, and Educational Experiences: What Should They Be to Address the Customer's Needs?
Session Chair: James M. Landwehr, AT&T Bell Laboratories

Educating Statisticians for the 21st Century 27
H. Jean Thiebaux, National Science Foundation

Modernizing Statistics PhD Programs 31
John Lehoczky, Carnegie Mellon University

Respondent
Joan B. Garfield, University of Minnesota 41

Respondent
Carl N. Morris, Harvard University . . . 47

Discussion . . . 52

PART II

Some International Perspectives . . . 61
S. Rao Jammalamadaka, University of California at Santa Barbara

Discussion . . . 67

How—and How Not—to Implement Content and Experiences

Session Chair: Daniel L. Solomon, North Carolina State University

A University Statistics Program Based on Quality Principles . . . 72
Edward D. Rothman, University of Michigan

Chance: A Course About Current *Chance* News . . . 86
J. Laurie Snell, Dartmouth College

Modernizing Graduate Programs in Statistics—Case Study . . . 91
Prem K. Goel, Ohio State University

Respondent
What Is Interdisciplinary Research? . . . 98
Stephen E. Fienberg, Carnegie Mellon University

Respondent
Ronald A. Thisted, University of Chicago . . . 110

Discussion . . . 117

Appendices

A Speakers and Session Chairs . . . 131

B Symposium Participants . . . 137

Opening Remarks

John R. Tucker
National Research Council

Welcome to this public symposium on modern interdisciplinary university statistics education (MIUSE). I am the National Research Council (NRC) program officer for the Committee on Applied and Theoretical Statistics, CATS, which organized this event.

The purpose of the MIUSE project is to identify the features needed in upper-level undergraduate and post-baccalaureate statistical sciences education to educate highly qualified professional statisticians for teaching, research, business, and industry, and to indicate how those features can best be incorporated into statistics programs. Support for this project has been provided by the National Science Foundation.

The idea for the project originated in 1991 in discussions by CATS about what it viewed as the most important needs in the statistics community. CATS believed strongly that a high priority should be put on completely rethinking the statistics major's educational program from the upper-undergraduate level through the postdoctoral experience. Subsequent discussions with statisticians on the NRC's Board on Mathematical Sciences and the Commission on Physical Sciences, Mathematics, and Applications, with prominent members of the statistics community, and with federal statisticians reinforced this view.

This symposium is intended to commence a national dialogue and examination within the nation's academic statistics communities of what modernization and change are needed. It is hoped that it will be a first step in what will necessarily be a gradual but essential period of discussion, experimentation, reevaluation, and feedback that leads to general consensus on what are the best ways to accomplish those changes that are needed, and what ought to be the minimum requirements for the diverse spectrum of modern statistics programs.

Ultimately, formal recommendations and a capstone report may result, but that is not the goal for today and tomorrow. This symposium starts the process by raising issues and concerns, providing a forum for the expression of ideas and possibilities, and sparking a nationwide examination through discussions of experiences and presentations of examples of what approaches could succeed in efforts to incorporate interdisciplinary training into upper-undergraduate, graduate, and postdoctoral programs, to bring the graduate curriculum up to date, to improve apprentice programs for graduate and postdoctoral students, and to reward faculty mentors for such activity.

PART I

What the Customer Needs in Statisticians

Session Chair: Jerome Sacks
National Institute of Statistical Sciences

The Content of Courses, and Educational Experiences: What Should They Be to Address the Customer's Needs?

Session Chair: James M. Landwehr
AT&T Bell Laboratories

What Industry Needs

Jon R. Kettenring
Bell Communications Research

Abstract

Industry needs holistic statisticians who are nimble problem solvers. They need to be able to work smoothly on teams and to communicate effectively about their work. Many of industry's statistics problems are interdisciplinary and involve exciting challenges in unplowed areas. Industry needs people who thrive on such opportunities.

Introduction

The industrial world continues to change at an amazing pace. In telecommunications, for example, it seems that the basic parameters of the business are in constant flux. The information age is here, but exactly what does this mean?

Along with the uncertainties are immensely challenging and exciting problems that need serious work, and much of this work is statistical in nature. Examples close to home include network reliability, software quality, and dealing with massive data sets. Generally speaking, there is a need to transform data into information (Nair and Pregibon, 1993) and intelligence that cuts across much of modern industrial life.

The major problems are usually of such complexity that progress is best made when they are tackled by cross-organizational interdisciplinary teams. Since many of these problems cry out for statistical thinking, it is natural for statisticians to be full partners on the teams. However, to be successful, statisticians need not only to bring their statistical expertise to the table but also to integrate themselves effectively with the rest of the team. Ultimately, the only thing that really matters from the industrial perspective is not the statistics per se but the impact of the total effort on the problem at hand. (See Duffy (1993) for a similar perspective.)

With this background in mind, it should be clear why industrial employers are particularly interested in hiring statisticians with a holistic view of the world as well as relevant and highly honed technical skills. Statisticians who can serve as bridge scientists (Lowrence, 1985, p. 7) are especially cherished.

The title of this symposium and the text of this paper reflect the presumption that a statistics education should be interdisciplinary. This concept is discussed in the next section, which is followed by an airing of the question of how a modern interdisciplinary statistics education can serve the needs of industry and by concluding comments.

Interdisciplinary Statistics Research and Education

Literature on interdisciplinary research is helpful for clarifying what is involved in interdisciplinary activities. Porter and Rossini (1986) distinguish this form of research, which involves substantial integration across disciplines, from multidisciplinary research, which is less integrated, and cross-disciplinary research, which implies only that two or more disciplines are involved.

The concept of interdisciplinary research is itself controversial. Caudill and Roberts (1951) point out numerous pitfalls. Seyle (1975, pp. 176-177) discusses how difficult it is to achieve interdepartmental scientific cooperation in the university. Bauer (1990) argues that interdisciplinary work involves surmounting cultural differences and is basically intractable.

However, there is no reason to be so pessimistic concerning statistics and its involvement with other disciplines. The starting point for the IMS Panel's report on cross-disciplinary research in the statistical sciences (IMS Panel, 1988) is the argument that "the driving force behind the development of modern statistics has been the need to solve practical problems." No doubt much of this work has been cross-disciplinary rather than interdisciplinary in the full integrated sense. Still, the emergence of the fields of biostatistics, econometrics, and psychometrics exemplifies involvement that has gone "all the way." Gnanadesikan (1990), while raising concerns about how vigorously cross-disciplinary opportunities in statistics are being pursued, points out opportunities in several areas, for example, molecular biology. The mission of the recently formed National Institute of Statistical Sciences is to provide a structure for cross-disciplinary research involving universities, industry, and government. In short, historical momentum, current opportunities, and some supportive organizational structures could all lead to acceleration of collaborative activities between statistics and other areas.

Within the university, there appears to be increased interest in building bridges between disciplines. Gardiner (1987) observes that leading universities are changing their emphasis from information gathering to information processing, which would certainly include a major role for statistics. He argues that interdisciplinary research groups, distinct from the usual discipline-based departments, are a vehicle for fostering such work.

Against this background, it is clear that there are, or can be, creative environments at the university that will support both cross-disciplinary or interdisciplinary research and educational activities in the field of statistics. Impediments, such as tenure concerns, are also recognized; recommendations for dealing with some of them are laid out in the IMS Committee's report on cross-disciplinary activities (IMS Committee, 1990).

The degree of integration across disciplines that is achieved in the educational process is probably less important than the commitment to teach statistics in a way that emphasizes teamwork, communication, and solving real problems. Students who have these values and associated skills ingrained in them are the most likely to be successful in modern industry.

What Needs To Be Done?

The potential for defining a successful holistic statistics education program—one that is right for the needs of industry, in particular—is within the grasp of most statistics departments.

The pieces are already in place in many cases, others are being experimented with, and still others have already been proposed. Departments that are unsure about how to proceed can obtain useful advice by surveying graduates who are working in industry. What may be most difficult is finding the commitment and resources to "put it all together." Since there is so much agreement on the need for change (for example, see Hogg, 1991), and those programs that lag behind are likely to find it increasingly difficult to attract and place good students, the chances of success would seem to be high.

Basic building blocks of any program should continue to be well-rounded statistical knowledge including subjects such as data analysis, statistical computing, sampling, linear models, experimental design, time series, multivariate analysis, and so forth. To make these courses really click, one suggestion is to make sure that they are interspersed with as much real data experience as possible (Singer and Willett, 1990; Cobb, 1991). A related suggestion is to assign instructors to these key courses who themselves have had substantial real experience with data (Gnanadesikan and Kettenring, 1988). Would a medical student want to learn surgery from professors who have never done it? These suggestions take on special urgency for courses on data analysis. For these courses, data-savvy instructors are essential!

Makuch et al. (1990) argue that an "optimal curriculum" for preparing statistics students to work in industry should consist of "topics from statistics, mathematics, computer science, and their applications." This is excellent generic advice. More attention needs to be given to correcting the imbalance between mathematics and computer science stemming from historical emphasis on the former. Makuch et al. go even further, suggesting that a graduate-level dual major in statistics and computer science would best prepare students for industry. The main point is clear: computing and related skills are vital for contemporary industrial — and probably all — statisticians.

Many statistics programs now have courses on and opportunities for consulting. Such activities are surely worthwhile, but they may not go nearly far enough in terms of giving students genuinely collaborative problem-solving experiences. In this sense, opportunities for students to work with interdisciplinary centers on the campus may be more fruitful and even lead to path-breaking cross-disciplinary thesis problems.

For some programs, closer ties with local industry may be a way of building an environment for experiential learning, a concept that has been emphasized by Snee (1993) and others. Opportunities for increased university-industry interactions seem to be wide open and can take many directions. Faculty and students both may be able to join problem-solving teams in industry. Industrial statisticians may be able to identify hot research topics for university statisticians to tackle. Experiences drawn from work with local industry should help faculty to enliven their courses. Whatever routes are traveled, increased contact between university and industrial statisticians should yield useful payoffs for both groups.

It is sometimes said that statistics students with subpar communications skills should take jobs in industry because they would not do well in the classroom. Unfortunately, they are unlikely to do well in industry either. In fact, there has been repeated discussion and concern about how serious a problem this is for industrial statisticians. McDonald (1988) emphasizes this in the context of communicating with management. Hoadley and Kettenring (1990) focus on the communications gap between statisticians and engineers or physical scientists, and explore many of its dimensions. Caulcutt (1987) raises similar issues from a British perspective. One is left with the impression that there has been very little attempt by statistics programs to respond to this very serious and legitimate concern. Statistics students must somehow learn to give an effective talk, to write a report that managers can understand, and to generally cultivate their interpersonal skills for statistical consulting and cross-disciplinary collaboration.

Conclusion

Industry needs—perhaps more than it ever has—nimble problem solvers with first-rate statistical skills. Solid statistical thinking is required for integrated solutions to many of industry's toughest problems. Statistical training should be truly holistic because that is the way statisticians are expected to operate in the real world.

To reach this state in university statistics education, closer working ties with industry would be worthwhile. This need not be painful. In fact, the results could well be renewed vigor between sectors that are sometimes unnaturally far apart.

Experiences along these lines should be shared so that progress can be made as quickly as possible. For example, Rutgers University has started a dialogue with statisticians in local industry about ways of working more closely together. What are they learning that would be of benefit to others?

The spirit of interdisciplinary work should shine through the education process. In both industry and the university, "as in farming, the most fertile soil may be that under the fences rather than at the center of long-established fields" (McHenry, 1977, p. ix).

Acknowledgments

Thanks to Ram Gnanadesikan, Innis Sande, and Vijay Nair for helpful comments.

References

Bauer, H. H. 1990. Barriers against interdisciplinarity: implications for studies of science, technology, and society (STS). *Science, Technology, and Human Values* **15**:105-119.

Caudill, W., and B. H. Roberts. 1951. Pitfalls in the organization of interdisciplinary research. *Human Organization* **10**:12-15.

Caulcutt, R. 1987. Statistics in industry: a failure of communication. *Statistician* **36**:555-560.

Cobb, G. 1991. Teaching statistics: more data, less lecturing. *Amstat News* **182**:1,4.

Duffy, D. E. 1993. Comment on "Is industrial statistics out of control?". *Stat. Sci.* **8**(4):380-384.

Gardiner, J. J. 1987. Building bridges between the disciplines: collaboration in an information processing university. *Eur. J. Eng. Educ.* **12**:121-127.

Gnanadesikan, R. 1990. Looking ahead: cross-disciplinary opportunities for statistics. *Am. Stat.* **44**:121-125.

Gnanadesikan, R., and J. R. Kettenring. 1988. Statistics teachers need experience with data. *Coll. Math. J.* **19**: 12-14.

Hoadley, A. B., and J. R. Kettenring. 1990. Communications between statisticians and engineers/physical scientists (with discussion). *Technometrics* **32**:243-274.

Hogg, R. V. 1991. Statistical education: improvements are badly needed. *Am. Stat.* **45**:342-343.

Institute of Mathematical Statistics (IMS) Committee. 1990. Report on Cross-Disciplinary Activities. *IMS Bull.* **19**(6):716-719.

Institute of Mathematical Statistics (IMS) Panel. 1988. *Cross-Disciplinary Research in the Statistical Sciences.* (Reprinted in *Stat. Sci.* **5**(February):121-146, 1990.)

Lowrence, W. W. 1985. *Modern Science and Human Values.* New York: Oxford University Press.

Makuch, W. M., G. J. Hahn, and W. T. Tucker. 1990. A statistical computing curriculum to meet industrial needs. *Am. Stat.* **44**:42-49.

McDonald, G. C. 1988. Communicating with management. *Chance* **1**:42-44.

McHenry, D. E. 1977. *Academic Departments: Problems, Variations, and Alternatives.* San Francisco: Jossey-Bass.

Nair, V. N., and D. Pregibon. 1993. Comment on "Is industrial statistics out of control?". *Stat. Sci.* **8**:391-394.

Porter, A. L., and F. A. Rossini. 1986. Multiskill research. *Knowledge: Creation, Diffusion, Utilization* **7**:219-246.

Seyle, H. 1975. *From Dream to Discovery.* New York: Arno Press.

Singer, J. D., and J. B. Willett. 1990. Improving the teaching of applied statistics: putting the data back into data analysis. *Am. Stat.* **44**:223-230.

Snee, R. D. 1993. What's missing in statistical education? *Am. Stat.* **47**:149-154.

What Academia Needs

Peter J. Bickel
University of California at Berkeley

I have been asked to speak on the topic "Modern Interdisciplinary University Statistics Education: What the Customer (Academia) Needs." My translation of this is: What kind of education would we like our future faculty to have so that they can best train the next generation of statisticians in industry, government, and academia? Or, equivalently, to what training and opportunities should we expose our own PhD students? I arrive at this by considering statistics departments as the customer for new statistics faculty.

As for "interdisciplinary," I think the adjective is redundant. Leslie Kish in his 1977 American Statistical Association presidential address (Kish, 1978) gives a description of our field that makes this point well:

> Statistics is a peculiar kind of enterprise of contradictory character because it is at the same time so special and go general. Statistics exists *only* at the interfaces of chance and empirical data. But it exists at *every* such interface, which I propose to be both necessary and sufficient for an activity to be properly called statistics. It has a special and proscribed function whenever and wherever empirical data are treated; in scientific research of any kind; in government, commerce, industry, and agriculture; in medicine, education, sports, and insurance; and so on for every human activity and every discipline.

Of course this interpretation is probably not what the organizers had in mind but rather the question, What do our colleagues in other departments want of us in terms of courses, consulting, and collaborative research? But my version is more fun for me at least, and I will necessarily touch on the other version also.

So what is the platonic ideal statistician? George Box's rhetorical description of Fisher (Box, 1976) serves well:

> "We may ask of Fisher
>
> Was he an applied statistician?
> Was he a mathematical statistician?
> Was he a data analyst?
> Was he a designer of investigations?
>
> It is surely because he was all of these that he was much more than the sum of the parts. He provides an example we can seek to follow."

The ideal is, as it has to be, far beyond our reach, but we can and should reasonably expect our PhDs to emerge with great strength in at least one of these categories and a serious acquaintance with all.

To achieve this, here is an unrealized and probably unrealizable graduate curriculum for the ideal statistics department. I list topics in alphabetical order because I believe they all should play some part in the education of a statistician. Which ones are emphasized necessarily depends on the inclination and talents of the individual, the range of talents and interests of the faculty, and the ultimate limitation—time.

1. Computing, including use of standard packages and development of skills in higher-level languages, use of simulation, statistical graphics, writing of usable software.

2. Data analysis, descriptive and exploratory, in the context of examples of substantive interest.

3. Data collection, handling, and quality.

4. Design of experiments.

5. History of statistics.

6. Inferences, both frequentist and Bayesian, model construction, testing, and validation.

7. Probability theory and probabilistic modeling in areas of substantive interest.

8. Statistical consulting and actual participation in interdisciplinary groups; breaching the language barrier.

9. Statistics in the law and public policy.

You may be surprised by my failure to include advanced mathematics courses in say, functional analysis, numerical analysis, discrete mathematics, and so forth. This is not to say that I question the truism that the more mathematics one knows the greater one's potential as a contributor to both theory and applications. Rather, I have tried to list topics with which I believe every PhD in statistics should have a serious acquaintance, wherever she is headed: academia, government, or industry.

Some of you may question the inclusion of statistics in the law and public policy. I can cite higher authority—Mosteller (1988) at p. 94 argues for ". . . a policy course taught in the regular curriculum for graduate statistics students"—and also personal experience at a recent planning meeting of the National Research Council for a further study on forensic use of DNA typing. The issues that arose involved questions of molecular biology, forensic science, population genetics, statistics, and the law—as interdisciplinary as you can get. It seemed to me of critical importance for the ultimate goal, good public policy, that the participants keep in mind the relative importance of the concerns of their fields and the interplay of these with the concerns of the other fields in the final outcome. I refer you to Mosteller (1988) for a subtle and extensive discussion of the need for policy study by statisticians. I would even argue, and I am glad to see that Jon Kettenring concurs, that holistic thinking is valuable in general in interdisciplinary studies rather than just for questions of public policy. Good scientific and technological enterprises have a goal, be it answering an important substantive question or developing a product, and it is this ultimate goal that statistician participants (and others) need to keep in mind.

Studying the history of statistics has several functions:

1. To be exhilarated to see that not only are things muddy now, they were even muddier then.

2. To recognize the highly multidisciplinary origins of our field in astronomy, genetics, agriculture, economics, engineering, and on and on. That is why I believe "interdisciplinary" is redundant when coupled to "statistics."

3. To see the emergence, submergence, and re-emergence of ideas as new types of data appear, as technology transfer occurs from one field to another, as old concerns are raised anew.

Stigler (1987) is a wonderful starting point, but we need more histories of the Fisherian era and the impact of the computer on our field.

You may have noted that in my references to data analysis and probabilistic modeling I referred to areas of substantive interest. By substantive I do not mean simply of interest to practitioners in some field. It is important to fire the imagination of our students by presenting methodology and analyses, in the context of data whose background is clear, that address well-defined questions that can broadly be perceived as important. At a supposedly elementary level Freedman et al. (1992) is full of such examples. At a more advanced but specialized level, *Fitting Equations to Data* by Daniel and Wood (1971) has such examples. The space shuttle failure analysis by Dalal et al. (1989) is another fine instance. The Boston housing data (Harrison and Rubinfeld, 1978), to me at least, does not fill the bill.

To what extent do we have such a curriculum at Berkeley? Bits and pieces at best. We have, I believe, excellent courses in statistical computing, general inference, high-level probability theory and a number of excellent "tool box" courses in applied statistics. My colleague David Freedman runs a course on critical analysis of historical papers of great substantive importance such as Snow on cholera (Snow, 1936) and Berkson on smoking and lung cancer (Berkson, 1955). We have a long-standing student consulting service that attracts graduate students and some faculty from many fields, particularly the biological and social sciences. These consultations sometimes result in longer-range interdisciplinary collaborations. A similar pattern is obtained through a summer placement program in industry. Finally, there are a number of ongoing placements of students in interdisciplinary collaborations in biology, engineering, astronomy, and so forth. These tend to come about through various haphazard faculty and student contacts.

Somehow it all does not hang together as well as it should. Certainly we have not built in requirements that ensure that all our students have some exposure in all nine aspects listed above. A major problem is time. All our courses are demanding. It is unreasonable to expect a student to take more than two to three a semester. And then the thesis looms. However, I believe there are solutions, for some of which we have examples. For instance, I would like to see courses divided into a number of shorter units, possibly taught by different instructors with perhaps some from industry and government with separate unit emphases on context presentation, modeling, data analysis, inference in this and similar contexts, and so on. Theses sometimes do, and certainly more can, have interacting general theory and application to substantive data parts. More mechanisms for facilitating interdisciplinary contacts of students and faculty can be constructed, and more are developing — for instance, the National Institute of Statistical Sciences.

To some extent, I find myself in the position of the comical Polish colonel in a movie some of you may remember, "Me and the Colonel." In this film the colonel and Jacobowsky (Danny Kaye) find themselves in a cafe in a French village during World War II carefully watched by the Gestapo who are only waiting to pick them up after they make contact with the Resistance. The colonel outlines their goals, eluding the Gestapo and getting to a nearby beach where an English submarine is to pick them up. Jacobowsky replies, "My dear colonel, this is all very fine, but how do you propose we do this?" The colonel turns to him in surprise and says, "But my dear Jacobowsky, I have outlined the strategy . . . the tactics I leave to you." I hope the rest of this meeting will provide the tactics.

References

Box, G. E. P. 1976. Science and statistics. *J. Am. Stat. Assoc.* **71**:798.

Berkson, J. 1955. The statistical study of association between smoking and lung cancer. *Mayo Clin. Proc.* **30**:319-348.

Dalal, S. R., E. B. Fowlkes, and B. Hoadly. 1989. Risk analysis of the space shuttle. *J. Am. Stat. Assoc.* **84**:945-957.

Daniel, C., and F. S. Wood. 1971. *Fitting Equations to Data*. New York: Wiley.

Freedman, D., R. Pisani, R. Purves, and A. Adhikari. 1992. *Statistics*. 2nd ed. San Francisco: W. W. Norton. 550 pp.

Harrison, D., and D. L. Rubinfeld. 1978. Hedonic prices and the demand for clean air. *J. Environ. Econ. Manage.* **5**:81-102.

Kish, L. 1978. Chance, statistics, and statisticians. *J. Am. Stat. Assoc.* **73**:1.

Mosteller, F. 1988. Broadening the scope of statistics and statistical education. *Am. Stat.* **42**:93-99.

Snow, J. 1936. *Snow on Cholera*. New York: Oxford University Press. (Reprinted, 1965, New York: Hafner Press.)

Stigler, S. M. 1987. *The History of Statistics (The Measurement of Uncertainty Before 1900)*. Cambridge, Mass: Belknap Press of Harvard University Press.

What the Government Needs

N. Phillip Ross
Environmental Protection Agency

I will talk a bit about what a statistician does in the federal government. Of course, the federal government involves a very broad spectrum of activities. The Environmental Protection Agency is a regulatory agency, and most of what EPA focuses on in the statistical area is public policy decision making.

By the Office of Management and Budget's definition, federal statistical agencies are in effect agencies that deal with public policy information. This means, in 80 or 90 percent of government activities, the idea of what statistics is for and what statisticians do differs from what I learned at the University of Maryland.

In recruiting, we look for statisticians that are persons for all seasons. We would like an applicant to be able to do everything and be a specialist in everything, not only a jack of all trades, but also a master of them all. But I doubt that being both a universalist and a specialist is possible. Nor is it possible to know everything and to become, for example, a universal chemist or physicist; yet you must become the expert, the person to whom everybody comes. Even in the policy arena, EPA looks for training and experience in working as part of a team, and especially so for a statistics PhD. An individual should not come into an organization thinking that he or she has to be an expert in every aspect of that field as well as in the substantive knowledge associated with the social and hard sciences needed to develop information systems, organize the data, and do the analysis for appropriate decision making.

Being on a panel as a client is a unique position for me. I am usually convening panels of my clients and asking them what it is that we should be providing them. Here I can instead say, "This is what I would like you to provide us," and I do not have to tell you how you are going to do it. As Peter Bickel said, I do not know how you would do it.

It is not necessarily the role of the university, in its training programs, its core curriculum, and its development of the degree, to provide this teamwork and communication ability. However, there are opportunities in the university program and process—cooperative agreements, internships, and other such things—so that such training does not have to be squeezed into classes. I teach a statistics tool course at American University, and I argue with the department that two semesters of it are needed rather than one, or rather than trying to squeeze more into the one semester. So I know first hand that finding time to do it all is already a difficult thing.

The most important thing I would like to see is people emerging from graduate school understanding that they are going to play on a team, and knowing how to communicate in that team setting. That is very difficult for people, especially those majoring in mathematics and statistics. As an example from my own experience, after working on my dissertation research for several months, it seemed rather trivial. I thought it certain that in my oral exams, the committee was going to ask, "Why did you bother to pick this problem; it is quite evident what the answer is." They did not. In fact, when considering teaching at universities and circulating the dissertation, at one university where the operations research department was interested in hiring a statistician, they sent my dissertation to the mathematics department because it was too complex for them to understand. And the mathematics department's response was, "We did not have enough time to go into it in depth, so perhaps Dr. Ross could talk about what he is doing right now as opposed to what is in his dissertation." As a recent PhD graduate, fresh out of school and with no experience, that was wonderful. They did not understand it and I did, and I thought it was trivial. The education that directed me, the competition that I had to go through as a graduate student, made me want to express something that I knew was right, but that others had difficulty understanding. It made me feel much better about myself.

When I went to work, initially in the private sector and then in the government, I remember talking with managers in a roundtable on an approach about which I said, "If you take that size sample, then you cannot infer." The person managing this roundtable said, "Hold it; speak English." Prior to that, I had not known that using "sample" and "infer" would pose a problem of understanding for some people. Although most people should understand that terminology, they do not for one reason or another. One of the critical things I have learned through experience in the federal government is to be careful about language, to be careful about jargon, and to become a member of a team that solves a problem as opposed to showing off my technical mastery, so to speak. Currently, statisticians who are hired straight out of the university must learn that on the job. Some do and some do not. As indicated earlier for industry, we at EPA also can no longer spend the time on that to acquire training on the job. Not only is industry cutting back in hiring; so also is the government. Having experience in teamwork is very important, especially with respect to communications. We want people who can communicate easily and across disciplines, and who will have the other team members understand what statisticians are trying to tell them.

Another desirable quality is an entrepreneurial element in the individual. This entails understanding that the problems you are going to solve do not come knocking at your door asking you to solve them. The people with whom you will work do not necessarily know what you can do for them, will not be coming out of their offices to ask you to solve some problem. In some government agencies, that did take place. I initially worked in the Army Research Institute, and we sat in our offices doing what we wanted until the telephone rang and somebody said, "We have a problem; come out and help us solve it." We did, and there was never a lack of work to be done. But in the public policy area, that is not the case. The managers and others are not necessarily sure of what you can provide them.

In particular, upper management in the government have not the foggiest idea what is going on below them in terms of the information needs that they have. Statisticians need to understand that. As an example of this at EPA, a few years back one of the assistant administrators for water gave a talk at a statistics conference. She was known for not having much use for statisticians, so we invited her to find out why. It was not a milieu of animosity or antagonism. She liked us, but just did not see what we did for her. One of her reasons she gave as an example. She had to advise the Army Corps of Engineers on whether or not they should be allowed to build a dam on a river relative to some environmental impact concerns. This advice had to be given to the Army Corps of Engineers in the next three months, and they would then act on it. Since statisticians obviously deal with data and provide information, she brought in statisticians from her staff and asked them to provide her with as much information as they could so she could make that decision. The answer from the statisticians was that the data that had been collected were not very good, the monitoring systems were set up to measure for compliance to the law, not with regard to the state of the environment or the impact that might occur. They said if they were given a couple of million dollars, and about two years, they would set up a survey design and monitoring system, and would come back and tell her what to tell the Army Corps of Engineers. This was not of much help to her, of course, because she had to make the decision in the next three months.

The students who intern at EPA do not understand that aspect of policy. They need this experience, to be confronted with this type of situation and individual. What is the role of statisticians in this policy office, and what should be their attributes? They should be people who would say, "Let us take a look at this; we will see what we can do." They should assemble a team of experts, because they are not the only experts, and do the best they can within the time constraints, explaining the limitations of the information, and cautioning the decision makers. Decision makers understand uncertainty. They do not necessarily understand it in the quantitative sense that statisticians may, but they do understand it.

The flip side of this argument is if you simply throw up your hands and say, "It is all garbage, you cannot do it, call somebody else," they will call somebody else; and that somebody else will do something for them and give them an answer. The next thing you know, the statisticians that said, "We cannot help you," start criticizing that answer. That is not a very good or popular way to operate in a public policy arena, and here is where statisticians run into trouble. Part of that trouble is that, outside of an academic world or outside of a research laboratory with which they might be affiliated in the university, they have had no experience or training in what their role would be in the real working world.

A statistician should be trained in a way to use the statistical thinking on the job, to be a problem solver, and not necessarily come to an organization or to a situation with a toolbox of methods and perhaps the mathematical training to innovate and modify those methods, but needing the problem to fit those tools. If I enter with a hammer as a tool, then everything looks like a nail and I make things conform to it. If my training in statistics is with experimental design using the general linear models, then every problem that I see will be framed by those general linear model tools in my solution attempts, instead of my coming into a situation where I am part of a multidisciplinary team in which we jointly look at what the problems are and try and figure out new and innovative ways to solve them. This difference does not rely on the necessity that I am a statistician and so have to solve the problem statistically. Surprising as it may seem, a good number of problems are not solved statistically. They have to be solved in other ways. But the way statisticians think and the way we train statisticians to think will ultimately help them to solve the problems.

The other aspect encountered in government that students must be able to deal with is the need to make a decision in three months for which a "quick and dirty" solution must be found. We have to start acknowledging that quick and dirty solutions will involve difficulty, but there are solutions and approaches using extant information and data that can be brought together to give indications of how to make the decision. That is actually all the decision maker wants.

As a client, I would thus like to see more students emerging from universities with experience using real data, but also with real data that are not very good. Very good data are not much of a challenge. Good data do not push the student to ponder what to do, but data that are not very good will. For example, the environmental monitoring and assessment program

(EMAP) is attempting to set up a probabilistically based monitoring system to provide data that can be aggregated to give a measure of the health of the nation's ecosystems. It is well thought out. The basic idea is to collect data in a statistical manner that would allow some form of inference and a relationship analysis to be done. However, in the federal government that is not done very easily because the Environmental Protection Agency is not the only federal agency collecting environmental data. The Department of Interior is implementing a new National Biological Survey in which they want to do the same thing. The Agriculture Department collects forestry data. The U.S. Geological Survey collects water data. The National Oceanic and Atmospheric Administration collects air and coastal information. Each of these agencies has designed its own systems, and each of these agencies has been collecting data for a long time.

EMAP wants to look at the whole picture. To do that, they have to use the data that the other agencies are collecting, not only the past data but also data that are being collected now, because those agencies are not going to reinvest hundreds of millions of dollars in their monitoring system to conform with the idea that EPA is proposing. So one of the real difficulties EPA now has is getting people who know how to deal with and integrate spatially and temporally different data sets, even with a known probability or sample distribution. But there is also the situation, such as in the Great Lakes or in the Chesapeake Bay, where the question is, How do you deal with temporally and spatially different data that were collected by different people at different times? This is an extension of, at least, the meta-analysis activity.

Statisticians must learn how to do that. It involves more than just developing methods. Statisticians must be developed that know how to attack those problems, people who can work with the other scientists to integrate and understand how to approach new questions and realize that the answers are not going to be perfect. That is what is really needed and is where teamwork and communication come in. There are interesting real problems of that nature requiring people to solve them. The nation has spent hundreds of millions of dollars on these data. One cannot just throw them out and start over. The Congress and others would not allow EPA to do so.

Those are some of the major characteristics I, as a client, would like to see in the statisticians who apply to EPA. I hope in this day-and-a-half symposium you can find ways to satisfy all these desires, because we would be delighted.

A Larger Perspective

John C. Bailar III
McGill University

As academic statisticians, we are missing the boat. We are barking up the wrong tree. We do not see what is plainly before us. We are kidding ourselves when we think that "our" kind of statistics is vital to the welfare of the nation and the world.

More and more, despite occasional appearances otherwise, we as academic statisticians are talking to ourselves. Even at this symposium we talk about how to do the old things better and more broadly, not about what we could offer to society, and what most needs to be done. Think about the whole range of the really big problems of the day: violence, crime and criminal justice, education and industrial productivity in the broadest senses, unemployment, the balance of trade, federal deficits, the health and welfare of millions of disadvantaged persons, urban rot, racial and ethnic tensions, and homelessness.

The kinds of statistics that we teach in undergraduate and especially in graduate programs have almost nothing to contribute to anything that matters on the scale of these problems. Instead, we teach about new abstractions in statistical theory, or we teach about new applications of theory to what are, in this context, tiny problems with tiny generalizations and tiny implications. I would certainly agree that every graduate student in statistics needs very substantial training in statistical theory, partly because of important applications, but partly also to recognize the many situations in which some theoretical development is not appropriate. In my experience, the latter has been more common than the former.

My own field, biostatistics, may illustrate a trend. Biostatistics began as an effort to break away from the sterile patterns of academic statistics, to develop and support practitioners who understand the problems and who contribute substantially to their solution. However, for many years biostatistics has been drifting back into this same kind of introspection, the navel gazing, the almost exclusive focus on theory, often theory that has no conceivable, important, real use. We fiddle while Rome burns.

Then we wonder why the world passes us by. Why is it that the economists, for example, are often on the front pages of our newspapers, or testifying to Congress, or making major decisions about public policy, but not the statisticians? Is it that we do nothing on that scale of importance, nothing that merits that kind of attention? I fear that the answer is "yes."

We teach what we enjoy teaching and what we know how to teach, not what the world needs. Think about that litany of problem areas I just recited. The solutions to those problems could profit enormously from sound statistical data, soundly analyzed. But the difficulties that block our understanding on these problems have little to do with probability models or random variation or all those other good things that make up statistics in academia today. They have to do, rather, with the vast range of other kinds of uncertainties, that is, what we broadly call bias. Bias dominates randomness almost everywhere. Think about your own past training and the training that many of you now deliver to new generations of students. What fraction of that training is or was devoted to bias? What fraction deals in any direct way with the big problems of this year?

When a "statistician" does take on the big problems and big programs, that person is very likely to be someone who has little or no formal statistical training. Ask yourself, for example, about the leadership of the big federal statistical agencies -- somebody mentioned a dozen; I might count as many as 15. These include, for example, the Bureau of the Census, the National Center for Health Statistics, and the Bureau of Labor Statistics. How many of these 15 agencies have as directors someone who claims statistics as his or her primary discipline? I think of just two. How many directors would benefit from a deep understanding of statistics at the highest levels? Is it less than 15? For many years, the person who held the title of Chief Statistician of the United States was not a statistician, but an economist. (I refer here to a prior incumbent; the person who now holds this office, also trained as an economist, is in fact a very accomplished statistician.)

I said that we talk about how we could do the old things better rather than how we might be doing the new things that need to be done. I discussed this matter briefly with Wayne Fuller of Iowa State University, who reminded me that if we go even further back in our history, our intellectual grandparents and great-grandparents in statistics did indeed devote themselves to these big problems, and good analysts found ways to use flawed and incomplete data to derive sound conclusions and practical recommendations.

But that tradition has been allowed to slip into the hands of other disciplines. For example, it has been epidemiologists rather than statisticians who have spent much effort in recent years on two areas critical to statistical analysis. One is understanding the nature of confounding and the effects of efforts to reduce its influence. The other is developing a taxonomy of bias. This taxonomy has some very important, big, practical implications. Work in both of these areas seems to be almost unknown to academic statisticians.

Again, it is other disciplines that have tackled the thorny issues of teaching persons who may be highly expert in some area of application but have the equivalent of grade school training in statistics about the use, sometimes even the correct use, of powerful computational and inferential tools. We have not wanted to get our hands dirty.

Jon Kettenring has mentioned here a set of five points that he considers important; I agree on all of them. One point was the need for data-savvy instructors. Three of the other four also rotate around this need for an understanding of what the world is truly all about. What do we do at present in academic settings to attract instructors who are data-savvy? What do we do to nurture them? What do we do to retain them when it comes time to consider tenure?

I have not said that current areas of great interest in statistical theory are unimportant or that our training is useless. Those areas are important and our training in those areas can be vital, but they cover only a small fraction of what needs to be done. For a moment now, let us think in top-down fashion. How many academic statisticians of the 1993 variety does this country really need for the two essential purposes of replicating themselves and producing a continuing flow of new theoretical results?

Then ask, How many academic statisticians do we have? It is my assessment that the present supply is already a lot greater than the demand; that the total professorate substantially exceeds what is needed for the really essential tasks of what is now academic statistics. Conversely, that professorate is far short of what we would need to take on the job I have outlined.

Are you with me so far? Do you think I am talking about your neighbor? I am talking about myself, and I am talking about you.

I have heard several questions asking speakers how they would solve problems that they have noted, and it would be quite fair to ask me the same questions. One thing we should do in the academic setting is to focus far more than at present on inference in the face of bias, sometimes serious bias. It was precisely this kind of concern that got me into the quantitative study of health risks due to hazards such as asbestos or dioxin (my current area of statistical interest, and it is statistics). In such risk assessment, uncertainties commonly range over three or more orders of magnitude. That is real uncertainty, and it is virtually all from bias. It is not going to matter much whether an animal experiment produces 20 versus 22 versus 25 cancers among 50 animals when you deal with that sort of uncertainty.

Another recommendation is to use big, real, important examples to teach theory. Any instructor who makes up a set of data to illustrate something has failed the students. If you cannot find a real example, there is something wrong with what you are trying to teach.

We are going to have to make cuts in the present curriculum. I am sorry; everything we now teach is surely important, but some things are more important than others, and we need to consider priorities. I do not think anybody else here has specifically brought out the need for dropping large parts of the things that we have traditionally taught, but there is not room for everything. Perhaps we all have a wish to teach our students everything we have learned over a period of, say, 30 years; it might be better to understand that they can and will continue to learn from their professional experiences, just as we did and still do. We really do not have to teach them everything we know.

We need to provide students with a substantially increased scope for electives. The curriculum that I recently inherited as chair of a department was chock-full, so the students had virtually no scope for electives, and that is one of the first things I am changing. Maybe you will look at your own curricula to see whether there is something that could be done about this, even to the point of allowing and encouraging graduate students in statistics to take more than occasional courses in substantive areas.

How about giving our students encouragement and support to spend a summer, or a term, or even a whole year in supervised work in some applied operation in government or industry? Some programs do this now, but not nearly enough.

My last point is that we must actively seek out and hire (and then give tenure to) statisticians who are especially good at doing and teaching applications. Will you give your next academic ticket to someone who has published half a dozen trivial papers in some annals, and will lead students who will do more of the same, or to someone who has used the tools of statistics to transform a national debate and who will lead students to do more of that?

Will these suggestions produce a different kind of graduate? Of course they will. That is the point of it all.

Discussion

CHARLES ANTONIAK: A question for Jon Kettenring: Is higher management at AT&T appreciative of these needs for statisticians in industry?

JON KETTENRING: As an employee of Bellcore, I will not attempt to speak for AT&T. However, I do not believe higher management generally has any particular interest in statisticians per se. They have considerable interest in people that they perceive as problem solvers and who can demonstrate through a track record of accomplishment that they contribute to the success of the business. Statisticians are as well equipped to make these contributions as anybody. So we have a special opportunity here, and in some sense we are negligent in not capturing this opportunity more effectively than we do already.

WILLIAM EDDY: Jon, where are we going to get space in our curriculum, in our years in school, to do all these things?

KETTENRING: I do not have an easy answer for that. One suggestion would be to integrate many of the things I described into the existing curriculum; that way you actually will get some synergy and hence some possible time savings. That may not do it all, but keep in mind that I need people that are as good at some of these things about which I spoke as the sort of people that you are producing for me today. And indeed, as you know, a lot of these people are not coming out ready for success in an environment where industry does not have time to retrain them. So if it forces statisticians to perhaps not go quite so far down some mathematical track, or so deeply into some more esoteric, theoretical statistical track, I am willing to make that compromise.

But I do believe that the experiences with real data, the experiences communicating, writing, and so on, can be integrated at least to some extent into an existing curriculum. I am not claiming it is easy. I believe this is going on in the engineering field in some schools these days, where they are trying to do some of the things I mentioned by integrating them into existing classes.

ANTONIAK: I taught at Berkeley for six years and then worked at Bell Labs for seven years, so I am familiar with both the academic and the industry environments. Peter, has Berkeley instituted anything in the way of Bayesian analysis?

PETER BICKEL: Of course, there has been a long-standing Bayesian course in the industrial engineering department. Berkeley also has a graduate Bayesian course that has been developed over the last couple of years by Andrew Gelman. When I mentioned inference, both Bayesian and frequentist, I meant it; both have to be taught as part of model-building. I devote time in my course to basic Bayesian inference, and I would like to expand it.

STEPHEN FIENBERG: Peter, where is there going to be space in your curriculum or in ours for these kinds of activities? And a related question: How do you bring these interdisciplinary activities into those theory courses that actually ask questions about inference? Is squeezing the curriculum to fit in these things going to be a workable mechanism?

BICKEL: I think it is. We at Berkeley currently have two parallel courses at the graduate level, one in theory and one in applications. This year David Freedman is teaching the applications course, and I am teaching the theory course. We are in fact talking about integrating the two in the future. It would be very difficult, and I certainly would hesitate to do that by myself precisely because I feel my own background is inadequate. On the other hand, there is enough expertise in the department to put together such courses. Whether all of my Berkeley colleagues would agree to do that is a different matter.

The time problem is critical, but a very exciting idea would be to work out ways of placing students in a coherent summer placement program in which it is clear that they truly learn some of the critical data analysis and communications skills that they need, and not merely have them be used as computer programmers. They could then bring back their impressions to the home department, provide some feedback, and perhaps some of their experiences could be incorporated into some of the courses.

EDDY: On your notion of dividing the courses so that we are sure to teach all of those activities: About 10 years ago we at Carnegie Mellon tried exactly that. We split all the introductory graduate program courses, and it was great because we certainly had a lot more courses and could teach many more things. But we found that each mini-course started to become a whole course, and the students began to collapse under the work load. In the last few years, we have migrated away from that and now have a very limited number of those divided mini-courses. Consequently, I would urge caution about dividing the courses and teaching responsibilities.

JEROME SACKS: One question that I have concerning the academic situation is the extent to which the environment within departments is suitable to get some of these things started. Peter, I believe you alluded to some difficulties with some of your colleagues. The question is, rather than just some of your colleagues, is it most of your colleagues or even all of your colleagues?

BICKEL: It is certainly not all of my colleagues. Either you or Jon said that things have to happen right away, that change has to happen immediately. I think, intrinsically, it cannot happen immediately just because the current faculty is a mix of people who were trained in different ways, some of which are more and some less attuned to interdisciplinary study. However, it seems clear that the way that the profession and the field are evolving, the great majority if not all of my younger colleagues would in fact support such initiatives.

MARJORIE HAHN: Peter, all of your comments were directed toward statistics PhD programs that are in separate statistics departments. Throughout the country, however, a large number of people get training in statistics within mathematics departments. Can you make any comments on that situation, or is the feeling that statisticians should be trained only within the statistics departments?

BICKEL: That is a very difficult question. We have in some sense faced that in our own department because we have a very strong pure probability group, and I believe there is much less willingness within that group to accept a view that a statistician should be the holistic "all-around-er" that Jon Kettenring described. I would hope that in mathematics departments the natural vehicle would be to form alliances with applied mathematics and to produce people who are PhDs in applied mathematics, because I think there is hope of having mathematicians who also recognize the need for some of these other skills. Also, there are going to be people who want to be pure probabilists, and I think that is fine. But as to what should be the typical product of a statistics department, I would like to see it be what I described.

ANTONIAK: Addressing public policy and the public's perception of what statistical measures are, I am sure that many people along the Missouri River, having suffered through the worst flood in a hundred years, now feel they are safe for another hundred years. There are other areas where there is still much controversy about the ozone layer and greenhouse warming. How do statisticians handle such tough issues and give guidance when there are large uncertainties?

N. PHILLIP ROSS: We do that on a case-by-case basis, exactly as you have phrased the question. We try to talk to people, mostly our own management chain. Statisticians rarely get called on the telephone by the general public. Sometimes contact comes through EPA's Office of Federal Activities, or a similar office. The answers we give are probably not very satisfactory, especially to someone who has experienced the actual incident for which we have given the probability of it happening. Although the incidence of an event is given "on the average," if you happen to be the one experiencing it, the average or the probability does not mean very much. If you are in a car accident, it does not help to be told that they happen only 1 in 10,000 times. But that is the only answer we can give.

J. LAURIE SNELL: As something of a philosophical question, to what extent should an academic institution ignore all the desires for particular training and continue being an ivory tower providing traditional education?

ROSS: As a person dealing with public policy, I believe universities have a role to play in the traditional sense. Everybody should be educated. My daughter has a friend who has decided to major in the classics, who, upon being asked the logical question, What are you going to do with that?, replies, "Teach the classics after I get my doctorate in it." I believe that in mathematics and statistics, universities should teach the theory, be the forum for debate, and provide students with those tools to prepare them not only to think but also to participate in the world of work, whatever that work might happen to be. In some instances it might be teaching; if students are to be well prepared for that situation, the art of teaching takes priority. There must also, however, be a way to expose students to real experience without universities inappropriately becoming institutes of vocational education.

There are people who wonder why they went to school and why they got the doctorate; it does not seem to have helped them at all. They wish that the university had told them there were not very many jobs in that field. There is some obligation to make students aware of such things, but I do not know how that is done in the curriculum, how to give a course for that. Integrating real experience, either outside of class or during the summer, is one way to address the issue. But the university has a role and is responsible for teaching the theory and raising the questions and the arguments, both those that may and those that may not be relevant in the real world.

CARL MORRIS: A small technical question for John Bailar: What do you mean by "bias?"

JOHN BAILAR: Consider the human cancer hazard of a specific level of exposure to a specific chemical compound, such as dioxin. There is great uncertainty attached to any number that can be produced from that. There are profound statistical questions involved with identifying the sources of uncertainty, correlations among those sources, producing estimates that are compounded, in a technical sense, with the total uncertainty. There is a need for a good bit of education of policy makers of all persuasions about the meaning of the uncertainty that is attached to these numbers and how to interpret it.

I commonly see something I call log thinking, that is, where someone thinks being a log off in one direction balances being a log off in the other. Of course, that is untrue. The difference is between uncertainty in an estimate of a specific biologic parameter in this context as opposed to true variation from person to person or subject to subject. All of these illustrate within one narrow field what I would like our students to know on a much broader scale.

Sometimes there are even opportunities to take bias and to study it in quite specific ways without being able to estimate a particular bias. Much has been said over the years about interviewer bias, and about bias in public surveys and other kinds of data-gathering activities. The questions ask how big the interviewer bias is. There is a well-recognized but not commonly used technique to study this. It has been used in the Current Population Survey and is called interpenetrated subsamples; it is able to produce a component of variance that can then be labeled bias. In the end one really has a probability distribution of biases. You do not know what the bias is for any specific interviewer. But I believe the concept can be extended.

DANIEL SOLOMON: I represent a large PhD-granting statistics department. We hired two new faculty this year. One has a PhD in physics, and the other has a PhD in genetics; neither has a PhD in statistics. Perhaps we should view that as a sad commentary that underscores Jon Kettenring's remarks.

In order to build a more interdisciplinary curriculum, the statistics community must encourage more interdisciplinary faculty. To do that, the reward structure for interdisciplinary work must be changed in the university. As long as publications outside the mainstream statistics literature are undervalued in the tenure process, faculty in the early stages of their careers are not going to be encouraged to undertake such work. Developing interdisciplinary work often takes longer than continuing work on one's dissertation, that is, continuing work in statistical theory, so that it is dangerous for new faculty to undertake interdisciplinary work.

Thus how is interdisciplinary work to be brought into the process? It falls on the senior faculty and administrators of the departments to continuously educate the tenure committees, the deans, and whoever makes decisions about tenure in the university, about the nature of statistics and the nature of interdisciplinary research. The reason this must be a continuous process of education is that those faces change and the educating must begin anew as new faces come on board.

BICKEL: This is a theme that has been sounded a great deal, and it has a great deal of truth to it. On the other hand, if you ask which way the arrow is pointing, it seems in fact to be pointing toward greater ease and rewards in that direction. To cite an example, Berkeley appointed a PhD in geophysics some years ago. I believe it benefits and enriches the field to bring in such people who eventually identify themselves as statisticians.

Concerning publications, I do not think that there are difficulties at the higher levels, at least at Berkeley, in persuading deans and tenure committees of the value of publications outside the field of statistics, because deans and tenure committees are usually not statisticians and would not know or care. The question is, What do the people in the department say about this work? What do the outside letters say about this work? There I think you may well be right that perhaps some of our colleagues still define excellence in rather narrow terms.

DICK BECKMAN: I am glad to see that universities are now appointing scientists in the statistics departments, because almost every young PhD these days is scientifically illiterate. Where I work, it is very hard for people who do not know anything about science to be hired.

SACKS: I have proposed to a number of people that the most important course in the statistics curriculum would be an integrated science course for graduate students in statistics covering some chemistry, physics, and biology.

CLIFFORD CLOGG: First a comment and then a question for any of the panel of speakers.

A great many of the statisticians that were role models in my own career, people who would be thought of as applied statisticians or methodologists who have made contributions to many different areas, have backgrounds that in reality were multi- or interdisciplinary. Some of these people had undergraduate training in psychology or engineering or computer science, and even graduate-level training that involved serious work outside a statistics department. It is striking that no one has mentioned strengthening multidisciplinary education; in fact, making statistics education multidisciplinary would mean having statistics graduate students do serious work in some other field. That does not mean that they take a "Bugs Bunny" course in statistics for economists at the 300 level, but rather that they do serious work in economics or serious work in biology or a serious project in urban affairs, and so forth. I do not concur on this umbrella course in integrated science, whatever that might be, because such courses tend to be a laughing stock for real scientists.

But is room available in the curriculum for this kind of serious work? Why not require that statisticians in the 1990s do some serious work in another field that gives them credentials and perspective, and so on, even if it extends the degree program an extra year?

BICKEL: That is something that has concerned me also. Some 10 years ago I unsuccessfully proposed in the Berkeley department that all of our students be expected to take a master's degree in another field, possibly mathematics. The difficulty, though, is that in the hard sciences, in the biological sciences, and perhaps less so in some of the social sciences, there simply is not time. Furthermore, our students just cannot do it. If you plunge a statistics student into a relatively high-level course in biology, he or she has to go back to the basic freshman course. I agree that doing serious work in another discipline is the ideal and I still believe that to some extent it can be done, but there are severe difficulties.

EDWARD ROTHMAN: I am troubled by the idea of sending students to learn a significant amount about one thing or another. There may be 20 or 30 courses that teach precisely the same principle. The conservation of momentum law will appear in a course in aerodynamics and in physics. We want to avoid that type of departmentalization. The physics department teaches $F = ma$, and the cardiologist talks about blood pressure and cardiac output, and so on, but basically there is a first-order push-pull law. What we do as statisticians is try to extract what is common to all these things. If we simply say to our students that they should gain expertise in a particular area, they will not grasp that big picture. I believe we have to provide that in our own courses.

BAILAR: We ought to do it ourselves, but the immediate implication is that we have to know how to do it, which means that we have to be doing it on a large scale ourselves.

ANTONIAK: Does that not also concern--and question--cross-disciplinary knowledge? As John Bailar says, a bias or unfamiliarity with the field leads to making inappropriate extrapolations of risk analysis of dioxin. Peter said that David Freedman has been teaching a class that covers good historical papers. Are there any classes reviewing great disasters, total misappropriation of funds, effort, and so forth, as in, say, asbestos or DDT? Students can get an idea of the pitfalls that are ever present when they get into a defined area. To name a classic example, storks bringing the babies comes to mind; the statistical tool is a correlation that we all think we understand.

BICKEL: I believe you will find a number of examples at that level in the Freedman et al. book I cited in my talk, but not major recent disasters, although I think it is an interesting question.

To slightly disagree, my son is starting a program in molecular biology at MIT in which one of the courses examines historical papers, but focusing on the points that are questionable: Was the evidence in the paper actually adequate for the conclusion that was reached?

Educating Statisticians for the 21st Century

H. Jean Thiebaux
National Science Foundation

Here I ask you to think of "the customer" as **the future of science** and to consider a case for a course of study in science, mathematics, and philosophy as offering the best undergraduate preparation for some future statisticians—including those who may find the most exciting work in the coming century.

It is important that statisticians be **partners in the future of science**—partners in setting and following agendas as well as analyzing the data fallout. This speaks to the benefit of statistical progression as well as to the benefit of other disciplines of science. Research in statistics and research in the companion areas of scientific enquiry are mutually supportive and progress together in their development.

Statistics may be thought of as **the art of persuading the universe to yield information about itself**. It provides the logical structure for scientific inference; for this reason, non-statisticians frequently think of statistics only in an immediate serving capacity. However, I want to entertain another point of view, namely, one that places the continuing development of our discipline as a major item on a bigger scientific agenda.

I will describe the developmental goal of statistics as the creation of a complete and unambiguous formalism—**a filter**—for taking in data received from the universe and giving back assignments of relative credibilities to possible truths about the universe. We can, and some of us do, work on this as a mathematical abstraction. However, it is intrinsically bound to developmental goals of other sciences. Optimally, statistics and its companion, probability, are full partners in the progression of scientific understanding. And what the customer needs from the statistician/probabilist is **refinement of the filter**. If we focus on a particular broad transect of current scientific activity, say, all the research dedicated to understanding and predicting global change, we can see that "refinement of the filter" means **extending its capacity to planetary-scale systems**, whether they are biological, geophysical, social, or political systems.

We ask and seek an answer to the question, What does this customer requirement mean for the education of future statisticians? History offers guideposts in seeking an answer, in the lives and work of the Bernoullis, Gauss, Galton, and Fisher. These prominent figures from the archives of statistics are all recognized for their considerable contributions to the creation of the filter. The patterns of their work support the conclusion that the future of science (as a customer of statistics) will be best served by statisticians who are solidly grounded in traditional science, mathematics, and philosophy.

We can trace the roots of our discipline back at least to the 1600s, to the Bernoulli family and three generations of scientist-mathematicians whose work encompassed physics, chemistry, astronomy, biology, and engineering, as well as mathematics. We are likely to know their name

NOTE: These are personal opinions of the author and do not represent National Science Foundation policy.

best, however, for their work in the theory of probability. My objective in citing them is to point to the breadth of their interests and acquaintance with science and to conjecture that the ground-breaking contributions the Bernoullis made in probability theory were inspired by questions that, during the more than 100 years of their work, were at the forefront of scientific enquiry (Bell, 1937).

In chronological order, the next on the list is Karl Friedrich Gauss, a prolific contributor to scientific research. Among his many accomplishments, this scientist-mathematician developed a basis for present-day multiple linear regression and spatial objective analysis. However, he did far more than lay a foundation for statistical analysis techniques that are based on what we now call the normal distribution. He is rightly named as one of the giants in the intellectual history of science. Gauss possessed extraordinary breadth of vision and insight, which he applied to the study of relationships among elements of deterministic systems. From these grew his formulations for estimating parameters from observations that are encumbered by uncertainty. In fact, the work that led to his formulation of least-squares estimation (in 1795) was directed toward the determination of planetary orbits, the scientific objective that inspired developments in statistical analysis of enormous scientific impact.

Francis Galton is next, for his description of the index of correlation in 1888 (Galton, 1890). Galton's work confronted puzzles of heredity, where these were confounded by influences of the nurturing environment. His discovery gave scientific understanding and mathematical expression to observed frequencies of biological characteristics (Stigler, 1989).

Finally we make note of the many contributions of R. A. Fisher, a man who studied quantum theory in combination with his studies of the theory of errors. From early in his education, Fisher too was seriously interested in the biological theory of evolution and genetics; and much of his work in the analysis of variance, for which he is best known, was done in direct connection with agricultural field experiments. See, for instance, Box (1987).

All of the above eminent figures in the history of statistics worked in scientific arenas, and their contributions to statistics were inspired by the need for structured inference in the sciences that fascinated them.

Consider, now, that the last of these major contributors worked at the beginning of the 20th century, and that there is a critical difference in the intellectual environments, then and now. Those who have been cited worked at times in the history of science when the perimeter of scientific understanding and formalism was much smaller. In those historic times, from 70 to 300 years ago, it was possible for one mind to be fully immersed in a process of scientific discovery and to know what was known about all its aspects. Scientific questions, and the intellectual energy of quests for their answers, inspired and nourished the earliest development of our field—by individual scientist-mathematicians. Although it is interesting to speculate about that relationship with scientific enquiry and tempting to envy its participants, it is no longer possible for a single mind to grasp the full complexity of many scientific questions.

From where we stand on the historic time line and look to the future, the perspective and opportunities are quite different. Now the "leading edge of discovery" is the edge of a much expanded scientific arena, following a century of technology-driven research. And now, many scientific projects are undertaken by research teams, and each research team is made up of several scientists. Team members will share scientific objectives and common scientific language, even through they may come from different disciplines. Thus the single mind of great vision has been replaced by a collective mind of suitable broad expertise.

What does a participating statistician need to bring to be a full partner, to further develop and apply the filter? Basic equipment should include a language of science and enough experience within scientific disciplines to appreciate and respect the methods and minds that are working to expand the knowledge bases of these disciplines.

Let us now turn directly to the topic of this symposium: "modern interdisciplinary university statistics education." As you might guess, I am going to rewrite that in terms appropriate to my customer, namely, as "education of future statisticians for interdisciplinary research." However, I am not going to presume to propose a curriculum, even for those future statisticians who will be scientific collaborators. Nonetheless, I do wish to raise some fresh possibilities for discussion. What the foregoing has to say about academic preparation is that we can do a great favor to students who believe they are headed toward interdisciplinary research by encouraging them to study science in parallel with mathematics, philosophy, and (maybe) a little statistics, in pregraduate years. The science content should best fit the student's true curiosity, and the mathematics should support future work on the filter.

For the "future of science" customer, with the statistician/probabilist as a full scientific partner, the complete education program will culminate in a PhD in statistics. Right at the top, that means that the student will complete a full complement of course requirements in statistical theory and methods in graduate school. With this as a "given," the one negative I am going to point out is my belief that the future of a creative, collaborative researcher is not optimally served by narrowing course concentration in undergraduate years solely to statistics courses. A broad perspective with solid grounding in the sciences will serve far better.

Now I am going to change tack a bit and support what I have said by anticipating what may be ahead for collaborative statisticians of the future. There will be the excitement of working at the boundary of science, developing probability and statistics in conjunction with investigations of our physical and biological habitat. And there will be energy and inspiration from sharing explorations of the frontiers of science:

- from gaining a scientific perspective,
- from accepting the challenge of representing physical and biological systems with mathematical abstraction, and
- from comparing its implications with subsequent measurements and observations.

These are privileged rewards of true partnership in interdisciplinary research.

Let me give this substance by describing a particular area. My choice is one that is receiving a lot of attention at present, because it concerns our ability to survive in our physically limited, global environment. It goes under the broad title of "global change research," and it involves a spectrum of critical elements of global sustainability. Among these are population distribution, marine and agricultural productivity, and energy requirements and resources.

Uncertainties in all areas of global change research stem from the limits of resolution of the information provided by global systems, whether they are geophysical, chemical, biological, or social/political. The scientific process of understanding such highly complex systems couples information from recent observations with established theory, for refinement of theory as a scientific goal, and for refinement of predictability as a societal goal. Thus, a key to reducing the uncertainties is to expand the limits of information resolution—**through the filter**.

The amount of time-dated data that has been collected and archived from Earth observing systems is truly immense, and the information contained in it is highly complex. Its volume and complexity magnify the challenge to data management and to the structure of scientific inference. In fact, the objectives of scientific inference should provide the framework for management of the data that is dedicated to scientific inquiry. Thus, the challenges fall within the forum where the scientific programs concerned with the environment come together with statistical science. Specifically, they are defined by the need for crafting filter algorithms that will output predictive estimates and statistical decisions with vast amounts of detailed (but noisy) input.

Classical statistical science has dealt with data sets that could be conceptually isolated. This is inadequate for estimation and inference in the larger framework of global change. High-resolution analyses aimed at reducing uncertainty in the inferences of global change research, which can be achieved through refinement of the filter, require bold new work. This work must take the true spatial and time coherencies of global systems as fundamental givens, and apply the philosophy of scientific inference in the creation of new statistical analysis techniques. Thus the requirement is for strong, focused, interdisciplinary teams dedicated to this mission.

This is a critical time in the history of the human use of our planet. Both scientifically and politically, the time is right to take on this task of refining the filter. There are inherent risks because the way ahead is unmapped, and big steps must be taken by collaborating scientists who have not traditionally worked closely together. The work will be challenging—at the conceptual limits of statistical science. The results can bring major refinements to the technology of assessment and predictability. The work will require statisticians who know both the language and the conceptual boundaries of the science.

Thus I conclude that students who wish to work on the filter in the context of 21st-century science need room in their academic programs

- to fully explore science,
- to achieve facility with applicable mathematics, and
- to contemplate the relationship between what we think we know and what we can measure.

References

Bell, E. T. 1937. *Men of Mathematics*. New York: Simon and Schuster.

Box, J. F. 1987. Guinness, Gosset, Fisher, and small samples. *Stat. Sci.* **2**:45-52.

Galton, F. 1890. Kinship and correlation. *N. Am. Rev.* **150**:419-431.

Stigler, S. M. 1989. Francis Galton's account of the invention of correlation. *Stat. Sci.* **4**:73-86.

Modernizing Statistics PhD Programs

John Lehoczky
Carnegie Mellon University

These are some thoughts on modernizing PhD programs in statistics to include more training in cross-disciplinary research. My intention is to provoke discussion by presenting some specific suggestions and raising issues that must be considered in any substantial overhaul of statistics curricula. I am focusing my remarks on the MS and PhD curriculum. Although the undergraduate curriculum is very important, there are so many differences between universities in terms of the student bodies they serve and the contexts of their programs that it is difficult to discuss undergraduate statistics education within a single framework. Nevertheless, if substantial cross-disciplinary training is introduced into the graduate curriculum, this will naturally offer opportunities for advanced undergraduate student involvement, and this will certainly enrich any undergraduate statistics program.

The Future University Environment

The question of modernizing statistics education cannot be addressed without first recognizing the significant forces that have taken shape in the last several years. These forces will have a profound effect on universities and their statistics departments. Some of these, such as the very difficult economic climate within which all institutions are forced to operate, are (it is hoped) only temporary. Others may be more long lived and can have important impacts, both positive and negative, on the field of statistics. They include:

1. A renewed emphasis on undergraduate education. New assistant professors should have some training in teaching, and current faculty must pay far greater attention to this aspect of their duties.

2. An increasingly negative connotation in the public's mind to "university research." This is seen as part of the decline in the education of students, and universities are seen as irresponsible institutions because of the very large increases in tuition costs and scandals associated with overhead recovery.[1]

3. Defense conversion and a strong national emphasis on economic competitiveness, health care cost reduction, the environment, efficient manufacturing, and so on. All of these

[1]The recent article of Peter J. Denning, "Designing new principles to sustain research in our universities" (P. J. Denning, 1993, *Commun. ACM* **36**(7):99-104), forcefully argues these points. Indeed, he addresses the modernization of the computer science curriculum, and his article offers useful parallels to those thinking about modernizing statistics curricula.

topics, and many others, are very large, cross-disciplinary problem domains that present tremendous opportunities for the statistics profession.

4. Industry's increasing attention to quality. Statistics is one of the important disciplines underlying quality control and quality improvement, and so statistics should be of increasing importance in engineering and business curricula. This represents a significant opportunity and responsibility for the statistics profession.

5. Industry's increasing need for technically trained employees who can be effective contributors to cross-disciplinary teams, and who also have strong oral and written communication skills.

A Major Dilemma

The focus of this conference is modernizing statistics education in the sense of fostering the inclusion of effective cross-disciplinary training. It is also clear that an increase in cross-disciplinary training can be of great benefit to the field of statistics. Through cross-disciplinary teamwork, statisticians can make important contributions to national problems, can identify problems that will expand statistics' core research agenda, and can increase statistics' audience and funding base. Thus it would seem that the only issue is to find and develop ways to expand cross-disciplinary training in the statistics PhD curricula. However, in thinking about how to foster cross-disciplinary training in statistics PhD programs, I was forced to think about all the material and tools our students must master to be effective statisticians, whether they participate in cross-disciplinary teams or not. As a graduate student 25 years ago, I always felt frustrated that there was an enormously large body of material that I had to learn, far more than could be learned over the course of a PhD program. The last 25 years have brought stunning developments that have greatly increased the amount of material our students must master in statistics and probability to be considered literate, not to mention the computer skills that must be mastered. Pull the latest issue of *JASA* or *The Annals of Statistics* from the bookshelf and scan the table of contents. It will show a whole new world of material unknown 25 years ago, using words such as "bootstrap," "Bayes-empirical Bayes," "CART," "projection pursuit," "ACE," "MARS," "Gibbs sampling," and so on, not to mention the renaissance of experimental design work spawned by new industrial statistics problems, or the tremendous growth of biostatistics both in applications and methodological advances. Statisticians are attacking problems an order of magnitude harder than those discussed 25 years ago; the wide interest in image restoration is an example. The developments are just as exciting (and equally daunting for graduate students) in probability theory. A student wishing to pursue probability theory must master texts on Brownian motion and martingales, and there is now a seemingly limitless amount of material to master on random fields, point processes, and so forth. The interesting aspect is that many of these advances have been spurred by applications. For example, the problems of image restoration, option pricing in finance, survival analysis, statistical mechanics, modeling large-scale distributed computer and communication networks, flexible manufacturing systems, and material fracture modeling have helped to bring about major advances in the many areas associated with probability theory.

There are two points: (1) cross-disciplinary studies have consistently led to a broadening and deepening of probability and statistics by posing new questions and bringing about new insights, and (2) the volume of material that a graduate student must master is ever increasing. While the first observation should harden the resolve to find formal mechanisms to help train statistics PhD students to effectively contribute to cross-disciplinary studies, never forget that there is an ever increasing body of material that students are expected to master to be literate statisticians and effective contributors to cross-disciplinary teams. The challenge is not only how to increase cross-disciplinary training, but also to find a way to do it that does not compromise the knowledge and skills a statistician must have to effectively contribute. In addition to the strong computing skills that are essential to being an effective statistician, students are now also expected to have the ability to make confident and articulate presentations, not to mention having good written communication skills. No one would doubt the advisability of possessing these skills, but the question is how to produce such super-statisticians in a finite (even relatively short) amount of time. I do not want to see statistics PhD training moving in the direction of complete specialization. It is common in other sciences for training to be very narrow, but this seems to me to be contrary to the spirit of cross-disciplinary collaborations.

CMU's PhD Program

I want to describe the Carnegie Mellon University PhD program in order to present some ideas of how we include cross-disciplinary training into our program, to point out strengths and weaknesses of our approach, and to highlight areas where difficulties might be encountered by departments wishing to develop this aspect of their PhD program. Before giving such a description, two points should be made.

1. In my view, true cross-disciplinary activities require the statistician to learn the language of the other discipline and understand the fundamental problems of that discipline. This is in contrast to working with a subject matter expert who translates the problem into the language of the statistician, often into a fairly precisely defined problem that is recognizable to (and perhaps solvable by) a typical well-trained statistician. I believe that the most important statistical skills in cross-disciplinary investigations involve structuring the questions to be asked and developing the methods of inquiry, as opposed to being able to pull an especially appropriate statistical procedure off the shelf. The important activities must be carried out with an understanding of the other discipline, its issues, and its methods, rather than in the language of statistics. For this reason, having each student undertake a consulting lab experience may or may not suffice, depending in part on whether such an experience results in a strong interaction with a subject matter expert over a sustained time period and leads to a good understanding of that field, its language, and its problems. Problems that are fairly precisely defined in the language of statistics do not lead to effective training for cross-disciplinary teams. It is also vital that there

be incentives for the student to invest the time and effort that will be required for a strong cross-disciplinary training experience.

2. At CMU, the faculty agree strongly about certain important aspects of PhD training. It is recognized that each student has individual strengths and weaknesses; however, every student who receives a PhD degree must have a solid mastery of basic probability and statistical theory, be knowledgeable about a wide range of statistical methods, be experienced at handling statistical applications, and be capable of computing effectively. The faculty is unified in its view that ALL students must meet the spirit of these requirements. There are no faculty arguments on whether to relax or even waive one of the standards because a student is so gifted in another area. All these abilities are necessary ingredients to functioning as a PhD-level statistician.

Students with a solid undergraduate preparation usually receive a PhD within 4 calendar years, including summers. The broad outline of the program follows.

Year 1 (MS Program)

Fall Semester

Perspectives in Statistics	(0.5 semester)
Statistical Computing	(0.5 semester)
Intermediate Statistical Inference	(1 semester)
Intermediate Probability	(1 semester)
Regression	(1 semester)

Spring Semester

Design of Experiments	(1 semester)
Discrete Multivariate Analysis	(0.5 semester)
Continuous Multivariate Analysis	(0.5 semester)
Time Series I, II	(1 semester)
Applied Bayesian Methods	(0.5 semester)
Statistical Practice	(0.5 semester)

Approximately 50 percent of the curriculum is theoretical and 50 percent focuses on the practice of statistics. The students must pass two exams, one that involves doing problems on the more theoretical material and one in which the students must demonstrate competency in data analysis. Some of the courses on applied subject matter involve applied problems that arise from faculty research. Also, students are expected to write reports. Among the courses listed above, two could be considered to offer some training in cross-disciplinary activity: Perspectives in Statistics, and Statistical Practice.

Perspectives in Statistics

The Perspectives in Statistics course lasts for seven weeks and is intended to serve as a relaxing, fun introduction to graduate student life in the CMU department. The initial sessions serve as orientations that introduce the students to each other, to the department, and to the facilities on campus. One session is spent on the "grading exercise," discussed in greater detail in the Training in Teaching section, below. The course also offers parallel sessions on computing, an introduction to CMU facilities, and an introduction to a series of standard statistical packages (Minitab, S, SAS, BMDP, and so on). The heart of the course is a series of one-hour lectures presented by each faculty member. The lectures allow the students to have contact with the entire faculty. More importantly, they allow each faculty member a chance to present a personal view of what is interesting in statistics. The only rule is that each lecture must be at a low enough technical level that it can be understood by this group of graduate students. About half of the lectures are on some identifiable area of statistics (for example, design of experiments, statistical graphics, or decision making), while the other half are on applications (for example, clinical trials, statistics and the law, the census, stock and option pricing, and Bayesian paleoethnobotany). This introduction provides important contexts that exemplify the department's broad view of the field of statistics and the role of statistics in many different kinds of problems.

Statistical Practice

The course on statistical practice introduces the students to consulting and interacting with clients. In the major activity, pairs of students work with a client on a real problem. For example, one pair might work with an analyst in the CMU planning office to develop insights into the undergraduate dropout rate, or why applicants decide to attend schools other than CMU.[2] There is a wealth of potential projects and willing subject matter participants, because the statistics faculty have many ongoing collaborations with other researchers at CMU or at the University of Pittsburgh. This course also covers material on report writing and presentation skills, and the students must make formal presentations. Thus, this course begins the teaching of cross-disciplinary research skills, but the course is far too short, and the project involvements are too brief and therefore too shallow.

---

[2]Statistics departments would do well to recognize that they can help to solve the problems that their own universities face. Many problems such as forecasting research revenues, student enrollments, and dropout rates are wonderful examples for projects. Statistics departments are uniquely positioned to offer such assistance, and successful contributions to important university problems will cause administrators to gain first-hand experience about the importance of statistics.

Year 2

Advanced Statistical Inference	(1 year)
Advanced Probability Theory	(1 year)
Advanced Data Analysis	(1 year)
Electives	(1 year)

The first two courses in advanced statistics and advanced probability are common to most PhD programs in statistics, with perhaps the only distinction being that the inference course usually contains material on the foundations of statistics and Bayesian theory. Students must pass qualifying examinations in both subjects. The advanced data analysis course is one in which the students tackle a true cross-disciplinary experience. The course has two major components that run throughout the year. One is a discussion of different types of data analysis problems, tools, and techniques. The students are assigned various topics (for example, ACE or MARS), read some relevant literature, especially examples of the use of the technique, and present the material to the class. The second, and perhaps most important, component consists of each student engaging in a major project involving an applied problem. The project is supervised by a three-person committee, one member of which is often a faculty member from some department other than statistics at CMU or the University of Pittsburgh. The faculty member in charge of the course also serves on the committee. The students write a major report, and all students make 30-minute presentations to the entire department on their projects. These presentations consume approximately two afternoons in May, and the faculty has been very pleased by the high level of achievement and the polished performances given by the students. Participation in a project over a six- to eight-month period provides a strong introduction to true cross-disciplinary activity, and discussions with other students who are doing different projects and having different experiences broaden the training. The committee arrangement facilitates the training in several ways. First, it widens the number of projects available to the students. If a single faculty member were to be in charge of all projects, they would likely all come from a single relatively narrow discipline. This way, the students get to indirectly experience other students' activities in unrelated fields. Second, it also obviously spreads the burden of this training across the department. Third, the wide faculty involvement in the advising and in attending the presentations reinforces to the students that faculty members are serious about this aspect of their training.

Years 3 and 4

The third and fourth years of the PhD program focus on dissertation research. Students select an advisor and begin reading and doing research in the summer after their second year. This results in a thesis proposal that is presented during the summer or early fall after their third year. Finally, the hope is that students will defend their dissertations in the summer of their fourth year. Students also have involvements in department research projects. Because our faculty have many cross-disciplinary research projects, this inevitably provides a variety of additional cross-disciplinary research experiences for the students.

During these last two years, students also take advanced courses. While a wide array of choices is offered, it is clear that these cannot be sufficient to ensure that the student is literate in all major aspects of statistics and probability. There is far too much material to cram into the remaining two years and still have time for the student to develop the research skills and research products associated with a good dissertation. There is a weakness in the lack of assurance that all students take advanced courses of a traditional sort in standard topics such as non-parametric statistics, multivariate analysis, sequential analysis, and so on. The tendency is to instead have students take courses that require them to read the literature, possibly in some focused area. The department does a fairly good job of discussing the foundations of statistics, and Teddy Seidenfeld offers a course on reading Fisher every other year that is greatly appreciated by all the students. In one very exciting advanced topics course that will be offered next fall, titled "Oldies but Goodies," two faculty members and the graduate students will be reading classic, landmark papers in statistics. There also are workshops in special areas (Bayesian statistics, biostatistics, and industrial statistics, for example) at which research problems are discussed in an informal atmosphere, and outside speakers are invited who present problems. In biostatistics, for example, researchers from the University of Pittsburgh Medical School frequently come to present problems, always preceded by a lecture on the relevant medical background. More recently, working groups consisting of faculty and students have sprung up for short periods of time to study some specific topic such as spatial statistics, Gibbs sampling, or graphics. As one who took a large number of advanced topics courses as a graduate student, I have always felt uncomfortable with the lack of required traditional advanced topics courses such as non-parametric or sequential analysis. Still, I believe that CMU students become very capable of learning on their own, which, after all, may be one of the most important skills a PhD statistician can possess.

Training in Teaching

Training in teaching and in written and oral communication is one final aspect of graduate training that is worth mentioning. This begins on the second day of Perspectives in Statistics with the grading exercise, and continues with more formal training and monitoring at the end of the first year. A few of our graduate students serve as recitation instructors for one relatively large undergraduate core course given during the year. True teaching experiences come during the summer in which all summer school offerings are taught by our graduate students. CMU has a very active teaching center, and the use of graduate students as instructors is very tightly controlled. All students whose first language is not English must be certified by our English as a Second Language (ESL) Center. Students must have received and passed teacher training. This entails mastering materials on course planning and syllabus construction, lecturing skills (including watching videotapes of good statistics graduate student and faculty lecturers), motivating students, grading and student evaluation, and so forth. The instructors are observed and videotaped, and they are critiqued by teaching center personnel on a confidential basis.

The grading exercise mentioned earlier is a device we use to try to achieve more uniform grading practices in our undergraduate courses and to encourage our graduate students to start thinking about teaching. The graduate students are given a copy of a fictitious exam, taken by a fictitious student in a standard undergraduate statistics course, and a solution key. Each student individually grades the entire exam according to an established point scale. In the next class, the results are compared. It is amazing to see the huge variability in the points assigned on each question and the overall scores. Overall grades range from B to D, the differences being due solely to grading practice. This variability is caused by different students having different models for assessing answers, including assigning partial credit for partially correct answers and dealing with irregularities. The variability is exacerbated by cultural differences and by the widely varying undergraduate educational experiences of these new graduate students. It was recognized that this variability had to be dramatically reduced, and it had to be done immediately, because nearly all of these students would be grading papers very shortly.

In addition to having completed teacher training, by the time each of our PhD students graduates she or he will have made many in-class presentations and at least three major departmental presentations, including the advanced data analysis project, the thesis proposal, and the thesis defense, each of which is attended by the entire faculty and student body. Even though the atmosphere is non-confrontational, the students generally find their first few presentations to be very stressful. It is felt, however, that after graduation, every student must be prepared to make presentations at conferences, to other professionals, or to management, and so these experiences are very important. Students are also encouraged to attend statistics conferences and present their research. All of this has been quite successful, and the department feels pride in the students' capabilities in this area.

Suggestions and Cautions

There are a number of suggestions, issues, and problems that must be addressed before PhD programs in statistics can include significant training in cross-disciplinary activity while simultaneously not shortchanging broad and deep statistical training. For example:

1. One must first recognize that graduate students face a sometimes overwhelming challenge to attain a PhD and to feel that they have a mastery of their field. Because this challenge is so great, the students often look to the faculty members for signs as to what is really important, what really counts, and what is really required. If a department is serious about ensuring that its graduates receive training in cross-disciplinary research, this message must be conveyed clearly and consistently to each student, by word and decision. Students will quickly see whether or not something is really important if they see full or sparse faculty participation, or see students who do poorly being required to undertake remediation or being passed in spite of apparent weakness.

2. It will be important to reach a strong consensus among faculty members on the importance of cross-disciplinary training, so that students receive a consistent story from the faculty. This will not be easy to achieve. The CMU department has evolved over time to the point that cross-disciplinary training is a significant component of our overall training. However, departments for which this is not the case may have difficulty in

establishing mechanisms for such training, providing incentives to faculty to lead its development, and reaching compromises about what current course work must be removed to make way for the new training. A faculty member will feel threatened if his or her special topic course moves from a required status to an optional status.

3. I believe that the responsibility for cross-disciplinary training should be widely shared among the faculty, not just placed in the hands of a small group interested in applications. This will be the hardest goal to achieve in most departments and will require senior faculty leadership. Moreover, if the faculty agree that this activity should be emphasized, it must be realized that some of them may not have any experience in cross-disciplinary work. Thus, these skills will have to be learned by some faculty as well as all graduate students.

4. The reward system in most departments must change so as to reflect the value of cross-disciplinary activities, both in promotion and tenure proceedings, and in year-to-year performance appraisals. Traditionally, this step is the most difficult for departments that have close associations with (or are administered by) mathematics departments. Even in autonomous departments of statistics, it will take time and leadership for the departmental culture to change so that cross-disciplinary contributions are thought of as being as important as papers published in our core theoretical journals. Departments will need to develop standards for evaluating coauthored papers and papers on substantive topics published in non-statistical journals.

5. The availability of computer facilities and statistical software and the ability of the faculty to integrate computing effectively into applications courses will be vital. This may pose difficulties for some departments that lack the necessary facilities or computer expertise.

6. Effective cross-disciplinary training requires access to interesting problems and subject matter experts who are willing to invest their time. For departments that have not established a large number of such collaborations, this will take a great deal of time and effort. Furthermore, an extra investment of faculty time will be needed to ensure that these collaborators are satisfied and willing to participate again.

7. Statistics departments might rethink the organization of their seminar series, to include outside speakers who will give broad overviews of topics rather than in-depth talks and to include more speakers on cross-disciplinary topics, especially in a workshop atmosphere.

8. The statistics community needs to continue efforts to attract survey articles on both theoretical and applied topics in statistics journals.

9. A centralized mechanism needs to be created for graduate student summer internships in industry, government, and medicine that will deepen students' training in cross-disciplinary activities.

Appendix

Examples of Statistical Practice Projects

- National Institute of Mental Health treatment of depression collaborative trial
- Longitudinal patterns of psychological distress following the Three Mile Island accident
- Factors related to early mortality among U.S. servicemen following deployment in Vietnam
- Discounts and quality premiums for illicit drugs
- Suicidal behavior in schizophrenics
- Analysis of clinical trial data on the effects of behavioral and pharmacological interventions in children with attention deficit disorder

Examples of Advanced Data Analysis Projects

- Analysis of data from a fiberglass production facility
- Determining the sources of lead contamination in soil
- Predicting enrollments at Carnegie Mellon University
- Survival times in patients with recurrent depression
- Analysis of oceanographic data
- A Bayesian analysis of bivariate survival data from a multicenter cooperative cancer clinical trial

Respondent

Joan B. Garfield
University of Minnesota

First, I want to share with you some of my reactions upon being invited to participate in this symposium. I thought there had been a mistake because, after all, I am not involved in a graduate statistics program, and I do not teach statistics courses to graduate or even upper-undergraduate majors in statistics. I am an educational psychologist whose area of research has focused on the teaching and learning of statistics, primarily at the introductory level. What could I possibly contribute to a discussion of modern interdisciplinary statistical education? However, John Tucker convinced me to give it a try and share with you some of my perspectives on teaching and learning statistics as they relate to the theme of this symposium. I must admit that I am glad he did convince me to accept his invitation, because I have had the opportunity to read some very interesting papers and find out that there are some very dedicated statistical educators who share some of my concerns and beliefs about educating students. I have already learned a great deal by reading the advance papers and listening to the presentations today, and will try to frame some of my reactions to the presentations we have heard in the context of my work on teaching and learning statistics.

Let me briefly summarize what I have heard so far: The purpose of this conference is to instigate change in upper-undergraduate, graduate, and postdoctorate statistics education. The focus is what changes in statistics education are needed to incorporate interdisciplinary training into these programs, bring curricula up to date, and improve the apprenticing of statistics graduate and postdoctoral students.

We began by listening to some excellent papers on what different customers are looking for in the statisticians they hire, hearing from the different areas of industry, academia, and government. Then we heard two papers on what the implications of these needs are for statistics education, and have learned about an exciting program at Carnegie Mellon that is trying to address these needs in some innovative and laudable ways. I heard some recurrent themes interwoven into these papers. I want to highlight a few, beginning with the needs perceived by the customers.

We heard about the need for:

1. Teamwork and collaboration: the need for statisticians to be able to work together, solving problems and working on projects, and the need for them to bring to these teams not only their statistical expertise, but also their knowledge of other disciplines involved so that they may contribute as full partners in the research effort.

2. Communication skills, both oral and written: the need for statisticians to be able to write and speak effectively about the methods and results of statistical analyses and the conclusions of projects undertaken.

3. Solving real problems with real data: the need for statisticians to be able to apply skills to a variety of contexts, know how to frame meaningful research questions, and help select appropriate methods.

These themes were echoed in the two papers that addressed what statistical education should be as viewed from academia and in a larger perspective. In addition, those presentations addressed two other themes:

4. The increased amount of knowledge to be learned (in the area of statistics as well as specialized knowledge in different disciplines, and particularly in science).

5. The need for internships and real-world experience in analyzing data and working on projects.

In the two papers just presented, we also heard much about what the content of statistics education should be to prepare future statisticians to meet the needs of our customers. Now, I wish to focus on the process of educating statisticians.

Keeping in mind the five themes I just outlined, I would like to share with you some findings from educational research that have implications for teaching and learning statistics. Then I will return to these five themes and relate them to the findings from educational research.

1. Learning is a constructive activity. Students learn by constructing knowledge and by being actively involved in learning activities. This contradicts the model of a student as an empty vessel or a blank slate, waiting to be filled with knowledge, as if knowledge is something that can be given or transmitted. Instead, the theory of constructivism describes learning as the process of integrating new information into students' previous knowledge and beliefs, as students construct their own representations of what is being learned. Much research in the areas of cognitive science, mathematics education, and science education supports this theory of learning.

2. Students learn to do well only what they practice doing. Research has shown that students become better problem solvers only if they have had frequent opportunities to solve problems. They become better communicators if they have had a great deal of practice trying to communicate their ideas and understanding.

3. Students learn to value what they know will be assessed. Students are astute at figuring out what they will be tested on and how they will be tested. Even if instructors profess to have other educational goals, such as learning to work together cooperatively or being able to understand important ideas, these will be taken seriously only if they also are included in assessment and grading.

4. Students seem to learn better when they are able to work in small groups on structured learning activities, when they are able to work on open-ended problems, when they are encouraged to write about what they have learned, when there is an emphasis on problem

solving and higher-order reasoning skills, and when they receive consistent and helpful feedback on their performance.

How do these suggestions relate to the themes mentioned previously? Students learn by constructing knowledge. Good students may be fairly adept at constructing knowledge from a lecture. By the time they have been accepted into a graduate program in statistics, they have already succeeded in undergraduate education, most of which consists of passively listening to lectures. However, students tend to learn better if they engage and struggle with material, rather than having it "delivered" to them. This works best when they are able to interact and discuss their learning with others. Not surprisingly, cooperative learning activities have been shown to be an effective way to learn, because students are actively engaged in constructing knowledge. If students develop experience learning to read and learn on their own or in collaborative situations, they come to realize that a lecture is not the only (nor the optimal) way to learn, and they should be more able to continue learning on their own outside of formal education. This is one way to deal with the theme mentioned earlier of too much content and too little time to learn it in the confines of an undergraduate or graduate program.

This leads us to the theme of teamwork and collaboration. Since students learn to do well only what they have practiced doing, teamwork and collaborative activities need to be an ongoing part of students' educational experience, and not just one isolated class in which students work on projects. Teamwork and collaboration should also be modeled by the sharing of experiences by faculty involved in cross disciplinary projects.

Encouraging collaboration and teamwork may not be an easy component to add to courses. Remember that students are used to being in competitive academic environments, where they are used to competing against each other, rather than working collaboratively with each other. Some may resist working on group projects, especially with group grades assigned. Students may be concerned about working with others who appear different from themselves. I would encourage you to consult some of the excellent literature on collaboration in higher education for suggestions and guidelines on how to successfully incorporate cooperative group activities into your classes (Artzt and Newman, 1990; Davidson, 1990; Garfield, 1993; Goodsell et al., 1992; Johnson et al., 1991; Weissglass, 1993).

The second theme I mentioned was communication skills. How do we know what our students have learned, and how do we know how well they are able to apply what they have learned? One important way to determine this is by giving them frequent opportunities to communicate their learning or performance, via written reports or oral presentations. By incorporating these types of activities into a class, we are actually doing several things:

- Collecting information to help us see what students have learned and how well they can apply their knowledge, and using this information to provide feedback about the quality of their learning and application;

- Giving students experience in learning to communicate effectively, by offering feedback on communication skills; and

- Demonstrating that communication is an important part of statistical work.

In addition, incorporating "writing to learn activities," in which students are asked to explain their understanding or interpretation of what they are learning, not only tends to improve their learning but also provides additional practice for students so that they may improve their written communication skills. Therefore, I believe that it is important to build written and oral communication skills into most of the classes students take, to help them develop these important skills needed by our customers, as well as to improve their own learning process.

The third theme I mentioned was solving real problems. We heard recommendations for courses to be built on real problems, real data, taught by people with experience analyzing real data, who can provide role models in data analysis. We agree that students need to experience solving or seeing problems solved in a variety of different disciplines and settings, involving resources or resource people from those disciplines. Again, these kinds of experiences can build collaboration and teamwork, give students practice in solving problems, and help them construct knowledge of *how* problems are solved in different contexts and disciplines.

I believe that students need to develop experience in solving open-ended problems, ones that may not be well defined and need to be clarified before they can be solved. In working on open-ended problems where a variety of approaches may be taken, based on different methods and assumptions, students should be encouraged to defend their selections, argue for their approach, and question other approaches and conclusions. These types of activities can help them develop higher-order thinking and reasoning skills as well as give them practice in problem solving.

If statistics educators agree that all of these different types of experiences are really valued, then we should make sure that students' educational programs include many experiences applying and improving their problem solving and communication skills and ability to work collaboratively with others. They should be assessed on their performance in each area and given appropriate feedback so that they may work toward improving their performance in each area.

This leads to a very different view of teaching, as the teacher's role becomes more that of a designer of activities, and facilitator of assessment and learning. This is a departure from the traditional role of teachers, viewed typically as "givers of knowledge," a view that is incompatible with the constructivist theory of learning.

Let me now attempt to summarize what I see as the implications of these issues for teaching statistics.

1. Statistics teaching can become more effective if teachers determine what it is they really want students to know and do as a result of their courses—such as the things we have been discussing today—and then provide course work, activities, and educational experiences designed to develop the performance they desire. The key here is knowing *what* students should be able to do as statisticians and making sure that students' educational experiences are aligned with these desired outcomes. It is interesting to note that the desired outcomes expressed by the speakers from the three different areas of industry, academia, and government did not specify particular content that students should know, but instead specified particular skills and experiences students should have.

2. Appropriate forms of assessment need to be incorporated into the learning process so that teachers and students can determine if these goals are being achieved.

3. Statistics educators need to take teaching seriously and assess their own personal theories of learning and teaching in light of the evidence classroom experience provides. If students are not developing the desired skills, or are not achieving the level of performance desired, teachers should rethink their notion of effective teaching and experiment with alternative methods, carefully evaluating the impact of these methods on student outcomes.

4. Students should be encouraged to assess their own learning and performance. An important goal of educating students should be to help them become better at assessing their own level of learning and performance, so that they may develop the appropriate standards for the discipline.

One aspect of being a good statistics teacher is modeling good teaching for the next generation of statistics teachers. One of the most exciting things I learned in reading John Lehoczky's paper was that graduate students in statistics at Carnegie Mellon University who serve as teaching assistants have to go through a comprehensive teacher training program.

I am inclined to believe that despite the different settings in which statisticians work, it is nevertheless important for them to be aware of good teaching and learning techniques, so that they may continue in their own lifelong learning, dealing with the continual increase of new information to be learned, and also so that they may teach others, whether in an academic setting with students or in another type of setting in government or industry. To this end, I think that course work and experience in teaching should be required of all graduate students in statistics, as well as for students in other disciplines.

However, I have some concerns about teacher training programs that focus exclusively on lesson plans, syllabi, handouts, and lectures. In keeping with the suggestions for good teaching I have described, I would like to see teacher training programs help graduate students learn about the teaching and learning process, learn how to develop and facilitate cooperative learning activities, become experienced with the role of assessment (and alternative forms of assessment), and learn about current ways to improve teaching in their discipline (that is, the use of software as a teaching tool, the use of projects, and so on). The development of programs such as these could lead to a new generation of improved statistics teachers and statisticians who are able to work more effectively in any type of setting. I look forward to seeing this happen.

References

Artzt, A., and C. Newman. 1990. *How to Use Cooperative Learning in the Mathematics Class*. Reston, Va.: National Council of Teachers of Mathematics. 73 pp.

Davidson, N., ed. 1990. *Cooperative Learning in Mathematics: A Handbook for Teachers*. Menlo Park, Calif.: Addison Wesley.

Garfield, J. 1993. Teaching statistics using small-group cooperative learning. *J. Stat. Educ.* **1**(1). [An electronic journal.]

Goodsell, A., M. Maher, and V. Tinto. 1992. *Collaborative Learning: A Sourcebook for Higher Education*. University Park, Pa.: National Center on Postsecondary Teaching, Learning and Assessment.

Johnson, D., R. Johnson, and K. Smith. 1991. *Cooperative Learning: Increasing College Faculty Instructional Productivity*. ASHE-ERIC Higher Education Report No. 4. Washington, D.C.: George Washington University.

Weissglass, J. 1993. Small-group learning. *Am. Math. Mon.* **100**(7):662-668.

Respondent

Carl N. Morris
Harvard University

Joan Garfield's comments are so important. I have tried to implement some of the things she said in my graduate classes, and I cannot do it very well. It has to do with the way we were taught. I was taught in the lecture mode, and I find it too easy to slip back into that same mode. I do not know if others here have also had this experience, but I will try again to follow Joan's advice because I think it is terribly important to help make that change. It will change the environment to a much more cooperative one.

I have resided in mathematics departments for much of my career although I have always been a statistician, and that will affect how I interpret things. I worry that we statisticians are running statistics a bit too much like the mathematicians do. However, I have also worked in other kinds of departments. At present I am half-time in a medical school, as well as half-time in the statistics department at Harvard. Before this, I was in the economics department at Rand [Corporation] for six years.

So my background includes having been both an employer, on the side of some of our earlier speakers, and a supplier of talent, in the view of Peter Bickel.

The most important thing teachers do is to decide on curriculum for their classes, to decide what it is that students are going to be taught. I believe this despite what Joan Garfield has said. I always feel that in the making of those decisions is the point when I, as a teacher, truly make a difference. And what all the presentations today have made evident is that some very tough choices must be made. We say that the graduate students in statistics today have to learn everything we did in the past, and that now they have to learn lots of new things, too. Furthermore, even though we have been in the field for more years than our students, few of us know these new things. But somehow the students are supposed to do all that. Thus curriculum choice is terribly important; that is what this symposium is about.

Who should be encouraged to enter the statistics profession? Should the historical mix of skills be changing? Should it move toward people who have other types of skills, sometimes sacrificing some of the more mathematical skills that are usually required right now? If that is the case, are the leading departments in statistics ready to do this? Moreover, what should be taught? Can the needed changes be made? I am pessimistic about this. It is awfully hard to change, because to do so requires performing surgery on ourselves, and that hurts. I will try to address a few of these topics.

Since I am a "respondent," I will review what preceded this. Jon Kettenring opened the symposium. He works at an institution that employs somewhat the same type of people as Rand does and did when I was there, and so I have some experience with that. Such institutions must emphasize interpersonal skills and recruit problem solvers. Those are necessary skills in interdisciplinary situations. At Rand, if people were hired who had good interpersonal skills and were problem solvers, and who were good at statistics, they would create new jobs. If you hire a statistician with the skills described by today's speakers, you will find that next year there will be a new opening created because somebody else wants a statistician who can help like that. On the other hand, when somebody without the interpersonal skills and natural inquisitiveness is hired, everything closes down for a while.

I am also suggesting that there is more than one dimension to a person's useful skills. Ability involves much more that just IQ; it is at least two-dimensional. When looking at the Graduate Record Exam scores for entering statistics students, the verbal should be considered much more than it has been in the past. I mean that both specifically, and also much more generally. A sense of the applicant's general skills must be obtained. This could change the mix of people who come in to statistics. It might also bring more balance to the profession, perhaps across gender lines.

Peter Bickel listed nine topics that should be covered in graduate training. One he did not mention—nor did anyone else--is almost the most important topic. That topic is model checking. I believe inferences are made by proposing a model and then checking it. Then you adjust it and check it some more. When you have done the best you can with your data and your knowledge, you draw the inference. This is conditional on the model you finally fit, the one you publish. Later, somebody else with better data, or with deeper understanding, might look again at the problem and show how to do a much better analysis. That is great; an evolution has occurred, and public knowledge has been enhanced. Both parties have made important contributions to this evolution. Certainly that point has to be emphasized when students are taught to do statistics.

What about training for the academic market, on which Peter Bickel touched? That involves training students to do something fundamental, and "important." It can be hard to understand the criteria. As a journal editor, I am occasionally told by authors that "this paper (or result) is very important." Sometimes I could not figure out what was important about it at all. I finally decided that when that happened, "important" meant that it was hard to get the result, and perhaps some smart people had tried earlier to get the answer and they could not. A recent example is Fermat's last theorem. Andrew Wiles recently established the truth of Fermat's conjecture (that there are no positive integer exponents n bigger than 2 such that A^n plus B^n equals C^n). The result will be hailed as very "important" in mathematics, because mathematicians have been working on it, unsuccessfully, for 350 years. But the world is not going to change much because of it, and I think this use of the term "important" is different from what would be understood by the non-academic market.

Often, however, even in statistics, the person whose genius establishes the statistical equivalent of the Wiles-Fermat Theorem still would be hired. Probably that is as it should be. But I am suggesting that more weight be given, in criteria for promotion and hiring of our faculty, to the needs for producing research and students that are important to the outside world.

Phillip Ross discussed expediency versus quality. The key sometimes is that the employer wants a fast answer, even if it is just a hunch and possibly not very good. That is acceptable practice so long as the statistician can say: "Here is a fast answer. It may not be very good, and I really need more time. We see this problem repeatedly, and because it will come up again, I must take some time to work on it and understand it for the next time." That is what statistics is about: getting important ideas that are used more than once, perhaps even in different fields, and abstracting them, developing them.

An illustration of this involves the widely used method of "regression" that was developed to learn how sons' heights regressed from fathers' heights toward the mean height. The word "regression" now refers to the method, rather than to the motivating application. I once had a similar experience with Brad Efron when we developed a method for shrinkage. We called it the "toxo" method, because we first tried it out on toxoplasmosis data. It was hard later to stop calling everything we did with that method the "toxo method."

I am trying to exemplify that in statistics we often do something to solve a special problem, but it really solves many other problems, too. That is something very worthwhile that statisticians do, and it is why statistics departments are so valuable. But statistics departments must communicate with other departments and with other research units to realize this value. Our statistics departments must do whatever is necessary to effect and preserve this communicating, using joint appointments, and so on, to nurture communication and interdisciplinary research.

John Bailar said that the supply right now exceeds the demand for theoretical statisticians. I agree. But I personally would rather not separate theory from application. I do not see how one can. I can think of few interesting applied problems that do not have theoretical components, or of any interesting statistical theory that has no applications. I abhor the sometimes-stated idea that one researcher is better than another because he is more "theoretical." Statistics students must be trained so as to be acquainted with both aspects, and to understand theory partly through applications, and application with theory.

Recently I have hit an impasse in trying to determine how to forecast future work load in Veterans Administration hospitals. Since I, thus far, cannot get the modeling of the data correct, I cannot get the estimates to come out right. This problem, which will have a big payoff, will take significant theoretical talent to solve. It is a theoretical problem, besides being an applied problem.

John Bailar also discussed bias. There are many important problems that can be looked at only if bias can be eliminated. For example, how can data be used to determine the relative benefits of one medical therapy versus another when the data were not randomized? There would be enormous amounts of data for doing that, if only the physicians applied the treatments at random.

I love Jean Thiebaux's idea of the collective scientific mind, that none of us is big enough to understand all the science but that somehow there still exists a collective mind that is the union of all of this. Perhaps some scientists do actually try to get an overview, just as a head of state cannot understand everything that is going on in the government but might have an overview of the government without knowing all the details. But I think the collective mind goes on within statistics, as well as across fields. Both Peter Bickel and John Lehoczky noted that graduate school cannot teach everything about statistics. So, there may have to be a collective statistics mind. Perhaps not every student can learn about sequential analysis. Instead, people will have to work together and to share knowledge. Thus Joan Garfield's cooperative role is appropriate in this context, too.

Ideally, I view statistics departments through a hub-and-spoke metaphor. In it, statistics sits at the center of the university, as the hub, and the other departments that use statistics are the spokes. The statistics department is in contact with every other department. Joint appointments help to do that, but there are many other ways to facilitate that communication, including communication through statistics students.

We learn from the other departments, and they learn from us. Statistical information flows between departments, often first through the statistics hub. The statisticians learn something from one department, they talk to other statisticians, and later a statistician who is working with yet another department spins the information back out, perhaps assisted by an abstraction of the methodology. Many mathematical ideas, such as regression, do not have to be rediscovered in every field. But somebody does have to communicate the idea.

This is why I believe statistics departments are needed; this is their principal role. However, we have sometimes abused our purpose (as mathematics departments, which should have a similar role, have done also). Departments of statistics (and perhaps even more so, our mathematics ancestors) are formed to play this role of contact hub, but once constructed, are tempted to change their mission to one of polishing up what was developed in the past. Instead, the need is to figure out what will be important in the future. Much of that can be discovered through interdisciplinary collaborations.

Concerning John Lehoczky's presentation, I have admired the Carnegie Mellon University graduate statistics program, and have thought that their department is probably the most successful in providing a first-rate statistics education. It is partly because they do the things that John described. They involve their students in many different ways in practice and in theory, and they help their students with communication and the interdisciplinary skills. But I do not see how it all gets done in the two years before the dissertation research commences.

John Lehoczky wondered what might be cut from the existing curriculum. I suggest that we cut many of those special topics courses, and perhaps have the student who is interested in specialization pick them up in the dissertation. This does not mean students cannot get course exposure to such topics. But there would be more learning such as what takes place in the first CMU course John described, where each faculty member lectures once or twice about his or her specialty. We need more courses that give people a feeling for the various topics. It is not necessary to teach everything; we should teach the most general ideas.

Students must be empowered to learn on their own. Joan Garfield touched on that. If it is done properly, they will learn to love the subject. They will be taught to learn on their own by being helped to find and read the literature. And, anyway, that is how we learn, by doing and applying.

It is going to be painful. Parents must forget their own egos as their children grow up. Our children do not have to follow our paths. They have to do what is right for them. There then will be evolution. We, as teachers, have to do the same thing, and teach what is needed. We have to let them do something different. Joan Garfield gave us some keys for that. If we cannot change our teaching styles, then we can at least let the students do more of the teaching, and they will change it for us. We need not replicate ourselves, to use John Bailar's term.

I would like to introduce a principle for statisticians. We must always do what benefits the field, and not our own department. We need to choose people who come into statistics to lead in the field, so as to help the whole field. We must not train students too narrowly, or the field will suffer.

I think the subject of statistics is extremely healthy. There will be ever more data, and data analysis, and computers to help in doing that analysis. When I fear that the field will suffer, I mean that statisticians will suffer. Many researchers do statistics. The question is, What will our role as statisticians be in the statistics enterprise?

As we consider the question of what hard trade-offs are acceptable, it might not be a bad idea to survey that portion of the statistical community membership that does the hiring. A questionnaire may help them to tell the people who are doing the teaching what they need to teach.

I was going to tell you a bit about the Harvard graduate statistics program. It is a small program with many good features that are quite consonant with the goals that we have laid out here. Only special students will do well at Harvard because students are left on their own a lot, and they do not take lots and lots of courses.

From the perspective of the general statistics department, again, the biggest step in making good curriculum decisions is that we first must make changes in our faculty selection decisions, and we need more joint appointments to facilitate all the things that we are saying. In some ways, smaller programs may find it easier to change than bigger ones because they more easily develop the unanimity that John Lehoczky said has been formed at Carnegie Mellon University. It is hard to agree on curriculum if people are all thinking differently, or if only one or two are thinking this is a good direction for the future. It is hard because there will be pain, loss of favorite courses, having to change research directions, and so on.

Another way all statisticians can affect research is to quit handing out the money for irrelevant projects. This can be done through the recommendations reviewers provide. When research proposals are being evaluated, one must not only evaluate how able the applicant is, but also how much benefit this research will do for the long-term health of the field of statistics. On the other hand, soft money to support applied projects can help a statistics department by keeping the department relevant.

Some of the science of mathematics surely will be preserved for statistics, partly because mathematical modeling is so important. I believe that 75 percent of what statistics does is create good mathematical models to describe the structures of other sciences.

The future of statistics is secure. Statistics departments, however, and statisticians might not be. We risk being ignored if we do not stay relevant. The statistics enterprise must not be allowed to fail because, as in the hub-and-spoke paradigm, it occupies a central, vital communication position that enhances the scientific capability of the entire university. People will truly appreciate what statisticians can do if we do what I know we can.

Discussion

ROTHMAN: I would like to hear just a few things about the reward system.

JOHN LEHOCZKY: You may not believe this, but in terms of value and cross-disciplinary research, at CMU we actually value applications papers, papers in journals that are not statistics journals, equally with—I would say more than—non-applications papers. And we value those just as much in our promotion, in our year-to-year performance appraisals, which have to do with year-to-year salaries, let alone promotion and tenure. We value those contributions in the same coin as contributions to *JASA* or the *Annals of Statistics* or whatever your favorite statistics journals may be.

We have those applications papers reviewed. If we ourselves cannot review them, because we cannot always judge such contributions, we have them reviewed by substantive experts to assure ourselves that they are in fact good contributions. We go through that extra step.

We do expect our faculty to have excellent credentials in some areas of the core discipline of statistics, whether it be Bayesian statistics or probability theory or time series, or whatever be the classic areas. The person has to have notoriety in some area or set of areas. The individual has to have credentials in statistics. And we want that person to be collaborating with other faculty members in the department, because that is a very important way our department works, and to be collaborating with subject matter experts in other departments in the university or outside the university.

So we value those applications and non-applications aspects in the same coin. Molly Hahn wondered earlier about mathematics departments, and I do, too. I think that for statistics departments that are within mathematics departments and whose faculty members are evaluated along with their partners within the mathematics department, achieving this will be incredibly difficult to ever pull off.

JEAN THIEBAUX: I did not hear John Lehoczky suggest that students with specific other disciplinary backgrounds be recruited to statistics graduate programs. Doing that is a different way of creating cross-disciplinary graduates, rather than retrofitting them. It does not take time from graduate concentration in statistics. Has CMU looked at that possibility?

LEHOCZKY: Our original master's program had as a concept that students could have a disciplinary area of their own, whether it be biology or oceanography, whatever the field would happen to be, and would study statistics. Our faculty strongly endorsed that as a concept. There is a unity in feeling that we are very interested in such students. But I think it is simply a failure of the recruiting process that we are not seeing those students. We are just not getting the applicants to have the opportunity to bring them into the program. And I think the failure is ours; it is a marketing question. But I agree wholeheartedly with the spirit.

EDDY: I am very interested in Joan Garfield's references to research in using collaborative work as a way to improve teaching. I am starting to teach a course in probability theory, and I want to try and use some collaborative learning methods that I have never tried before.

JOAN GARFIELD: A paper of mine that just appeared in the new, first issue of the *Journal of Statistics Education*, an electronic journal available on gopher, is specifically on using cooperative learning in teaching statistics. Concerning the research that says this is a more effective way of learning, some of my colleagues at the University of Minnesota, David and Roger Johnson, have put together a huge literature review [see p. 46, above]; I believe they cite over 250 articles that have shown that students do tend to learn better, that is, achievement seems to be higher, when they work in groups.

ANTONIAK: We technical types tend to be more impressed personally with these computer-generated animated demonstrations of principles. We tend to think that the right way to get a concept across is to find a new, good way of demonstrating something, by being colorful. What comes to my mind is Tom Apostle's work with Project Mathematics in which, for example, there are very neat demonstrations of the theorem of Pythagoras. Joan Garfield's presentation focused mainly on the methodology, the dynamics, the learning environment, the interpersonal dynamics between the students and the teacher. Have any studies been made on whether the cleverness of the demonstration is 30 percent, or 50 percent or whatever, of what is required? Or are you really saying that the biggest problem is in recognizing a different way to go about teaching any kind of material of this type?

GARFIELD: I am not sure I understood the question. I first thought you were going to ask me if there was research on the use of computer graphics, demonstrations, and so on, and would that have an impact on student learning? Is that part of what you were asking me?

ANTONIAK: Yes, basically.

GARFIELD: I think that studies are starting to be done on the use of different kinds of software and ways that students interact with them, and what is the most effective way to use that kind of software to help students learn. And it seems to me that it is very encouraging. With technology offering such sophisticated ways to demonstrate things that we have never been able to present before, it seems to offer the potential to help students understand very complex concepts in better ways than they had previously. But I cannot say that there is a set of literature out there that supports that right now.

JOHN TUCKER: Computer technology can be very effective in improving student learning, especially in reinforcing class presentations and for self-paced instruction, but its effectiveness strongly depends on how well or poorly the software is designed.

GARFIELD: Right, and how well students are able to interact with it, whether it is just a demonstration or whether it permits them to manipulate variables.

ROTHMAN: I would like to address this issue of grading and assessment and also encouraging cooperative learning at the same time. If in fact we put people in competition for a grade, do we not undermine the purpose of learning?

GARFIELD: I think I have a different view of assessment than that. My idea of assessment is giving feedback to students on their learning, not just handing them an end-of-the-term grade. It is more an ongoing interaction with the student that says, here are some areas of weakness, here are things I think you need to work on. I view assessment as an ongoing process, and as a very complex process where ideally we would be giving feedback to students on their statistical knowledge, how well they apply it, how well they communicate it, and so on.

I think that assessment is very much a part of collaborative activity because if a group works together and turns in a product, they need feedback on how well they did on that product. I know that most professors view assessment as grading, and I think that issues do come up when you are grading group work and students are worried about their grades. There have been different suggestions in the literature on different approaches to dealing with that.

ROTHMAN: Specifically, how do you feel about comparison between students? If you base assessment on how well they present their work, you are making a comparison, relative to someone else rather than to what that student has already done.

GARFIELD: I guess I do not see assessment as comparison to other students. I see it more as comparison to a standard: "Here is what we would like you to be able to do, and you are not there yet, but here are some suggestions for areas you should work on." I personally do not think of assessment as a way of comparing students to each other. I do not do rankings. I believe in more of a mastery approach, whereby if every student in the class masters things to the level I am looking for, they will all get the same grades.

JAMES ROSENBERGER: The idea that statistics is at the hub of a hub-and-spoke paradigm is one of the themes that I have encountered here that Pennsylvania State University is probably not aware of. I wonder if we need to do a great selling job of the statistics discipline for the rest of the academic community.

MORRIS: I am afraid so. But let us get started.

FIENBERG: One of the problems that is going to come up repeatedly, and has been alluded to everywhere, is how to fit everything in. In reflecting, I have been associated in one form or another with at least five different departments over my career, and every department has had this problem. So it was not a problem that only I encountered; indeed, it existed at Harvard when I was associated with that university early in my career.

John Lehoczky was correct in saying there is clear agreement on the goals and the importance of data analytic and cross-disciplinary training. But it is also very clear to me that there is not unanimity at CMU about the curricular details. Further, one could probably put any pair of people together who, when looked at from afar could seem to coincide, and find they think very differently about the curriculum.

A number of years ago, when I was at Minnesota—before Joan was there--I observed that, when put together in a room, the faculty demanded the union of the knowledge of all of the people in the room rather than the intersection. A consequence is that you add course requirements and you never take them away. If allowed to go to its ultimate end, you have an infinite-year curriculum, a curriculum that cannot work. So there is a serious problem here.

The other observation comes from my life as an administrator, which is now over; I am now languishing back into the field of being a faculty member. We in statistics are not alone. We talk about this as if it were unique to statistics, but in fact every field in every university faces these decisions. In fact, the pace of curricular reform and change is similar in other places, and indeed, I believe statistics in many respects is moving more rapidly ahead. In my most recent administrative role as a vice president, I was astonished at the slowness of some fields' willingness to embrace the notion that you had to reexamine what you were doing, let alone change it.

What I would commend to everybody is to think not in terms of 2 or 4 or 5 or even 10 years as the increment for comparison, but to think in terms of generations and centuries. If you go back a century, statistics did not exist as a discipline. If you look at universities a hundred years ago, there was not an English department because it did not exist as a separate, identifiable field. And therefore, anybody who tells you that you cannot change the curriculum over that length of time is just talking from ignorance. If you use that long-term view, you know that change has to occur, and the question is how rapidly you make it happen, and how acceptable it is to be making changes regularly. Statistics as a field has actually been a good model for that. The notion of process control, where you do make regular changes and adapt, is something that we have been teaching others for years. Perhaps it is appropriate to take that and bring it back and use it ourselves as we adapt.

MORRIS: It is always easier to make change when departments are being built. The first statistics departments in the United States were formed just before World War II. Changing is much harder once you become institutionalized. Statistics is going to be more like the classics department the next time.

SACKS: Joan Garfield offered a set of tactics to go with the strategy that had been discussed before by Peter Bickel. Do you have, or do you know if there has been attempted, an assessment of the cost, in terms of time or resources, of implementing those sorts of tactics at a graduate level, or even at an undergraduate level, or whether it is cost-effective to do so?

GARFIELD: I do not know how to answer that. I have lots of suggestions on how it can be done, but I am not directly involved in a statistics program, and so I cannot speak to what the cost would be.

ROTHMAN: At the University of Michigan, we have one class in which we use masteries or portfolios rather than tests at the end of the period of time. Students demonstrate their understanding by writing something that indicates that they understand the facts and that they can apply the facts to situations that have not been described in class. They have to go to a newspaper, a scientific journal, and say, "Here is an application of this principle to some other situation." They get some feedback from the teaching fellow, and then they either have mastered the topic or have to revisit it. So the grade is "mastered" or "not yet." That simple change from assigning numbers as grades is very important because it focuses on learning as opposed to performance on tests. We are out there trying to encourage learning.

We have a class of 250 students. Even being involved in this teaching college for more than one term in this section, using three plus two other graders, it is a full-time job just getting involved in that seemingly small change. We are trying to find new ways of doing it by putting more of a burden on the student, and getting some software that allows them to check their own work. We have the $4,000 to do that this summer and will see how that plays out. We are going to have to do a lot of work to get the cost down.

The bottom line is that, from my understanding, it is going to be a very expensive policy.

EDDY: What we have been talking about seems to me to be the distinction between educating students and training them in something specific, and that the historical mode of lectures is to drum the information into them. What these various things that have been talked about this afternoon really focus on is educating students so they have the tools and savvy for these situations, and not worrying so much about training in the specifics.

In thinking about this, I am still struggling with my earlier question of how we are going to teach them all of these things. The answer is that we are not, and we do not have to worry so much about it.

Also, John Lehoczky omitted mentioning one of the other mechanisms that we at CMU have incorporated in the last few years, namely, small groups of students and faculty members that get together. We now have six or seven of these groups that meet two or three nights a week, in which the students have to make presentations. It is so much smaller; it is not a course or anything of the like, it is just a get-together or workshop. In the ones in which I play a role, in the course of a month the students probably make one or two presentations. They get feedback on the communications part and on the technical part, and their fellow students get exposed to whatever ideas they are discussing. So there are other mechanisms that are imparting the knowledge that they did not get in course work, and in realizing this I actually feel much better about it than I did earlier.

JAMES LANDWEHR: I want to comment about this issue of how one covers more in the same amount of time, from the perspectives of having worked with the Quantitative Literacy Project and of trying to get high school and middle school teachers to teach more statistics. Of course, if you say to a mathematics teacher that, in addition to everything else, he or she now should be teaching statistics and the students should understand statistics by the time they graduate, the immediate response is, "Well, what can I throw out? You tell me what to throw out, what is not important, and I will consider it."

Eventually other teachers may say to them something like the following: If we think about it differently, if instead of spending a lot of time teaching them linear algebra -- in which the straight line is presented so abstractly that the students do not get it anyway and more time is thus taken than should be — we give the students some real data and ask them what it means and talk about scatter plots and look at association and eventually ask the class, "Could a straight line help us in understanding this?", we may end up not only teaching some statistics, but also teaching the simple algebra better — and I am now repeating what teachers tell me — so that the kids end up learning the simple algebra about the straight line better than previously happened the traditional way.

So if you can come up with a different method, sometimes you can kill two birds with one stone. That is not to say it is easy, but I think this is a perspective worth taking, as opposed to the "What-can-I-get-rid-of?" perspective.

BICKEL: I have to agree completely with Steve Fienberg that when you get the department together, the tendency, of course, is always toward the conclusion that one must offer the union of all things rather than the intersection. Furthermore, it is driven not only by the faculty interests that lead to that union, but also by what skills are, say, most relevant in the environment. For universities in the Washington, D.C., area, focusing on survey sampling and statistics policy may be most relevant; for the University of Michigan, focusing on Teach UM and its various aspects may be what is most important. On the other hand, it is clear that is impossible to offer everything. So there has to be selection. Nothing prevents there being a large number of topics available as long as you can get the faculty to agree on the intersection, that is, on what every student should have some exposure to. Then have the other things offered, or offered as time permits or as there are faculty members willing to teach them.

BODAPATI GANDHI: There has been discussion about cost. We try to extend the basic philosophy of quantitative literacy to undergraduate students who are taking undergraduate introductory courses. We ask them to do a project instead of a third partial examination. Each student has to collect primary and secondary data, prepare a proposal, a small one, and then analyze the data on a computer. It is done at no cost to the university. There have been about 10 sections with each section containing about 34 students, so that around 300 students total have experienced this for two semesters. In the first semester they focus on descriptive statistics, and in the second semester they do a regression analysis project. We offered it this year also and it was very successful, according to comments of the faculty.

SNELL: I was just going to comment that we at Dartmouth and many other people have experimented, and many have approaches that are very expensive, and also some other methods are very cheap. I heard a wonderful presentation the other day about a physics program at Harvard in which the person who teaches a group of a couple hundred people simply comes in each day with a very short question, gives the students a few minutes to think of the answer, and has them turn around and discuss the problem with a neighbor and come to a joint conclusion. This is done merely to exhibit by first-hand experience how much better they do after they have talked about a problem a little bit with each other. He happened to have a lot of high-technology equipment that probably was expensive, because all the statistical results were displayed in front of the room almost instantly with automatic recording and such things, to also show the students how much they had learned from just one or two minutes of discussion with their neighbors.

However, to do something such as that is very simple and does not cost a lot of money.

PART II

How—and How Not—to Implement Content and Experiences

Session Chair: Daniel L. Solomon
North Carolina State University

Some International Perspectives

S. Rao Jammalamadaka
University of California at Santa Barbara

My aim is to share with you some thoughts on statistical education and, specifically, bring in some international perspectives. Before I do this, I should tell you a bit about my background. My entire statistical education (bachelor's, master's, and the PhD degrees) was at the Indian Statistical Institute, Calcutta, under such people as C. R. Rao, P. C. Mahalanobis, and J. B. S. Haldane. I have been teaching for the past 25 years, mostly in the United States, but I have also had opportunities to teach and observe statistical education at various universities and institutions in India, Sweden, and Australia.

There is no argument that statistics is a very interdisciplinary subject. Statistics derives its strength, usefulness, and its research motivation from applications and the practical problems that arise in other fields. Therefore, I believe that, for them to become effective statisticians, statistics students need to be trained not only in statistics but also in one or more substantive disciplines. To this end, I advocate the idea that statistics should be offered at the undergraduate level, exclusively or mostly as a "double major," in conjunction, say, with computer science, economics, biology, environmental science, sociology, psychology, and so forth. In this connection, I will describe in the next section some of my own experiences as a student.

At the master's and PhD levels, students still need to learn "theory" alongside the "practical" methods. The question as to how much of each, the "balance" issue, depends on what these students are training for. Given a finite amount of time and a finite number of courses that students can be asked to take, and the almost unlimited variety of topics available, we have to be very selective in what we teach. I shall briefly describe in the second section below our experiences in this context at the University of California, Santa Barbara (UCSB).

Statistics — Theory versus Applications

I believe in the theme that statistics is a "technology," much like medicine or engineering, and not an art or a science. Mahalanobis (1965) distinguishes science as the "effort to know nature more adequately" (that is, learn the truth—often for its own sake), and technology as "the effort to use scientific knowledge for the fulfillment of specific purposes." He also says that a technologist, unlike a scientist, must have a "knowledge and experience of a wide range of scientific subjects," and he quotes Fisher: "a professional statistician, as a technologist, must talk the language of both the theoreticians and the practitioners."

NOTE: A substantially similar version of this paper was published in the *Bulletin of the International Statistical Institute* **54**, Book 1, 3/7, 1/9. This updated version is reprinted here with the permission of the ISI and the author.

Fisher, Mahalanobis, and Haldane were, in the main, responsible for formulating the degree programs and courses that have been initiated and taught since 1960 at the Indian Statistical Institute, Calcutta. This institute is truly a university with statistics at the center—a concept to which Carl Morris alluded earlier. I was among the first group of students selected for the 4-year professional statistics degree program, the B. Stat. (as contrasted to the 3 years it normally takes to complete a bachelor's degree in India). This very ambitious program's objective was "to offer comprehensive instruction in the theory and practice of statistics and provide at the same time a general education, together with the necessary background knowledge in the basic natural and social sciences, expected of a professional statistician. Students were taught a multitude of subjects where statistics finds applications. They included biology, economics, physics, chemistry, genetics, sociology, engineering, demography, and so on. Practical aspects of statistics were emphasized just as much as the theory. For a more detailed description of this curriculum, see Rao (1969).

Even before we students knew what statistics was about, we were taking part in a post-census survey for the 1960 decennial census and trying to understand such basic notions as "household," "family income," and how crucial such an understanding is for the survey to be meaningful. One of the class projects consisted of extrapolating the 1960 census figure from all the previous censuses and the post-census survey data. One biology project under Haldane, of which I have a vivid recollection, was to count and record the number of petals on hundreds of flowers of a particular species, so as to learn not only the botanical aspects of these flowers but also statistical concepts including frequency distributions, variation in nature, and so forth. Later, after learning a bit about sampling, we were involved (knee-deep in mud, literally and figuratively) in a "crop estimation" survey. The goal of these projects was to prepare statisticians who could analyze data sets and also have a good understanding of the context.

This bachelor's degree program demanded that, in addition to mathematics and statistics, the student learn a great deal about the wide variety of substantive areas. While the program's goal of training statisticians with a broad understanding of the applications was laudable, I believe it was far too ambitious. What now remains at the Indian Statistical Institute is but a skeleton of this idealized experiment at training well-rounded statisticians.

Although the master's degree program was very mathematical, my PhD thesis research tackled a problem in paleocurrent analysis, involving directional data, brought up by a geologist at the Indian Statistical Institute. When this geologist later inquired about a sampling strategy for the paleocurrent data, I was sent to the "field," a very inaccessible forest where he was collecting such data, so that I could more thoroughly understand the context in advising him.

However, from the foregoing description of the Indian Statistical Institute of the 1960s, which I consider to be ideal training for statisticians, I do not want you to get the impression that statistical education is being optimally managed at all Indian universities. Unlike the Indian Statistical Institute, most Indian universities suffer, as do most places, from lack of resources, both financial and intellectual.

For U.S. universities, I strongly advocate a more modest scale of incorporating one or more application areas at the undergraduate level—say, as a "double major" in statistics along with another substantive area such as biology or economics. In a bachelor's degree program, students should spend approximately equal amounts of time on the four subjects of mathematics, computing, statistics, and application areas.

At the master's degree level, it should be possible to separate the students into at least two streams, depending on their goals and interests. For instance, a "Mathematical Statistics" stream would be for those who wish to go on to a PhD, and an "Applied Statistics" stream for those who wish to make the master's a terminal degree and join government or industry. These latter students should be exposed to a wide range of statistical techniques and methods, such as applied multivariate analysis, categorical data analysis, some biostatistics including survival analysis, sampling methods, and time series. They should also be able to effectively use statistical packages for data analysis, be able as consultants to solve general statistical problems from industry and government, and be able to communicate with other scientists. There is an excellent report by the Statistical Education Section of the American Statistical Association on this type of training (ASA, 1980). At UCSB, we found that students who obtain their BA or BS degree in one of the substantive areas (for example, biology or psychology) are equally successful at pursuing this stream, if they are willing to make up deficiencies. This "retro-training" of substantive discipline majors for graduate study in statistics ties in closely with my recommendation of double majors.

Students in the more theoretical stream should take additional courses in mathematics, probability, and statistical theory, as appropriate to their chosen research topic. However, I firmly believe that students in the PhD degree programs should be encouraged to develop considerable breadth in statistics and familiarity with applications and computing, before they are allowed to specialize in their chosen research topic. This might be achieved, for example, through various course requirements or "qualifying examinations." To obtain a PhD in statistics, a student should have working knowledge at the basic level of, in addition to computing, various areas of statistics such as survey sampling, design of experiments, and time series. Students should not be awarded the doctoral degree in statistics for abstract exercises in "statistical mathematics" having no connection or even potential applications to analysis of real data.

Students at every level of statistics education should get the sense that the subject is driven by applications. The best place to teach such applications, whether in an elementary or an advanced course, is alongside relevant theory. Teachers do not always find this to be easy or convenient. Most feel comfortable with tossing coins and dice for "concrete" examples. It seems that the training (or retraining) should actually begin at the faculty level, with the faculty developing interests in some application areas. This can be achieved through joint appointments or by statistics faculty going on temporary assignments to government or private agencies, so as to acquaint themselves with real-world problems. Other ways for statisticians to keep in touch with reality include participation in a statistical consulting laboratory at the university, and/or private consulting. The statistical laboratory can also be an educational tool for the students and an extremely useful resource for the rest of the university.

Whether statistics is a separate department or resides in the mathematics department, building bridges between statisticians and the substantive disciplines, which sometimes teach their own courses on statistics and have a slightly different agenda, can be a rather delicate task in most U.S. universities. It is crucial that statistics departments find common ground with them in order to build both joint educational and joint research programs. At UCSB, such interaction is encouraged by "adjunct" appointments being provided in our group for those faculty from the substantive disciplines with statistical interests. If some become frustrated in making such efforts, consider that in Sweden, the split between applied and mathematical statistics groups is somewhat more formalized and traditional than in the United States. There, many universities have two separate departments with no apparent interaction. I strongly believe that such a state of affairs is detrimental to both the groups.

The Program at UCSB

At UCSB, the Department of Statistics and Applied Probability currently offers programs for (1) a BA and BS degree in statistical science; (2) an MA in statistics with three possible specializations: applied statistics, mathematical statistics, and operations research; and (3) the PhD in statistics.

The two bachelor's degree programs require a substantial amount of course work in mathematics, computing, and statistics. However, at present, the programs do not require course work in application areas such as I recommended above. We hope to introduce several double majors, where, for instance, statistics majors will also study biology, economics, or computer science. The BA degree requires 10 one-quarter courses in statistics (30 fifty-minute sessions of instruction) at the upper-division level, while the BS degree requires 13 such one-quarter courses. An abbreviated, categorized list of courses (all of which are one quarter unless otherwise indicated) is as follows:

Requirements:	Probability and Mathematical Statistics (3 quarters) Design and Analysis of Experiments Regression Analysis Statistical Computing
Options:	Sampling Techniques Statistics in Industry Sequential Methods Nonparametric Methods Operations Research Ranking and Selection Methods Applied Stochastic Processes (2 quarters; includes some time-series) Actuarial Statistics and Risk Theory (3 quarters)
Others:	Internship in Statistics Independent Studies in Statistics

The Independent Studies in Statistics course is a vehicle for learning any other topic that is not taught as a regular course. The Internship in Statistics encourages students to participate in a faculty-supervised, academic internship in industrial or research firms in the area. This is an excellent way for the students as well as faculty to relate the academic course work to real-world problems. Faculty who supervise such interns and their projects should receive appropriate teaching credits, if not other considerations and rewards, when they are evaluated for merit raises and promotions; however, we have not yet been able to arrange for that at

UCSB. A reduced teaching load is given to the faculty member who runs our statistics consulting laboratory.

I should also mention in this connection the very successful annual "Careers Day" held by the Southern California Chapter of the American Statistical Association, in which our students participate. This is an excellent way for students to learn about what jobs are available and what kinds of course work prospective employers look for. This annual "Careers Day" is now being duplicated in Sweden by the Swedish Statistical Society's Educational Committee.

An MA degree with any of the three specializations requires about 11 one-quarter courses (mostly graduate level, but with a few approved upper-division undergraduate courses allowed). An abbreviated list of currently offered graduate courses (all of which are three quarters unless indicated otherwise) includes:

Statistical Theory
Multivariate Analysis
Statistical Decision Theory
Linear Models
Life Testing and Reliability (1 quarter)
Operations Research
Case Studies in Operations Research (1 quarter)
Advanced Statistical Methods
Seminars and Projects in Statistical Consulting (1 or 2 quarters)
Probability Theory
Advanced Probability Theory
Stochastic Processes
Advanced Stochastic Processes

There is also a catchall course called Seminars in Probability and Statistics, in which topics (generally advanced) vary with the instructor's interests.

Students in the Applied Statistics master's program take the Advanced Statistical Methods course that covers a wide range of topics (not always with formal proofs), and they have to do an internship in the statistics consulting laboratory for a quarter or two. Here they participate in actual consulting under supervision and write a project report on their statistical analysis and conclusions. This is where the students are exposed to the "inconvenient" real-world problems that may violate traditional assumptions, may have missing data, and may indeed call for the development of new theory. The Operations Research specialization requires a similar case-study project after a year-long graduate course in operations research has been taken. Students in the Mathematical Statistics specialization take the more theory-oriented courses (Statistical Theory, Probability Theory, and so on). These students are encouraged to keep up their interest in an application area by academic credit being provided for appropriate courses taken outside the statistics department.

To make sure that the PhD students have the appropriate breadth, they are asked to take "qualifying examinations" in any four of the following six areas: mathematical analysis, mathematical statistics, applied statistics, operations research, stochastic processes, and probability theory. There are still two foreign-language requirements, and one of them can be fulfilled by demonstrating proficiency in a computer language. Almost all the students take this option and thereby do a rather substantial exercise in computing.

Concluding Remarks

Although there is no unique formula that works for training all categories of statisticians, I believe there is general agreement that statistical training as well as statistical research should be driven by applications. We statisticians are in the fortunate position of being able to motivate all the theory we teach through real-world applications, and this synergy between theory and applications should be conveyed to our students. A good understanding of the theory along with its limitations, an appreciation for applications, and a familiarity with statistical computing should form the basic themes for any university degree in statistics. To this end, I recommend the idea of double majors at the undergraduate level, the establishment of an active statistical consulting laboratory, internships in statistics, and coordination with the local industry and government.

Since statistics is a discipline that is still evolving and changing, it is imperative that the courses and curricula be reexamined on a periodic basis and updated as necessary. For instance, the tremendous power of the computer now at our disposal has revolutionary implications for statistics and should be taken advantage of. Computer-intensive methods such as resampling techniques (including bootstrapping), iteratively reweighted least squares, density estimation, simulations, dynamic and interactive graphics, and so forth have all become topics of great interest. See, for instance, Speed (1985) and Tierney (1990). These topics should be incorporated as well as a host of new and emerging theoretical areas such as inference for stochastic processes, robustness and influence functions, directional data analysis, spatial statistics, image reconstruction, interacting particle systems, and so on, into statistics curricula alongside the more traditional courses that are already offered.

References

American Statistical Association (ASA). 1980. Preparing statisticians for careers in industry: Report of the ASA Section on Statistical Education, Committee on Training Statisticians for Industry. *Am. Stat.* **34**:65-67.

Mahalanobis, P. 1965. Statistics as a key technology. *Am. Stat.* **19**:45-46.

Rao, C. R. 1969. A multidisciplinary approach to teaching statistics and probability. *Sankhya, Ser. B* **31**:321-340.

Speed, T. P. 1985. Teaching statistics at the university level: How computers can help us find realistic models for real data and reasonably assess their reliability. Pp. 184-195 in *Teaching Statistics in the Computer Age*. L. Råde and T. Speed, eds. Lund, Sweden: Studentlitteratur.

Tierney, L. 1990. *LISP-STAT: An Object-Oriented Environment for Statistical Computing and Dynamic Graphics*. New York: Wiley.

Discussion

MARY GRAY: While I found the discussion of the students' experiences in India to be interesting, university education really is a very delicate issue from an international perspective. In particular, most statistics departments in this country are training a very large number of people who will be working internationally. Unless we believe that we are just exploiting a brain drain of statisticians from around the world, the people that we are training are going to go back and work in quite a different atmosphere, and there has not really been any consideration given to that. What the customer needs when the customer is Bellcore may be quite different from what the customer needs in some other countries of the world, particularly the countries of the south. Notwithstanding the particular experience described in India, it is my experience that the split between the theoretical and the applied is much greater in most countries--particularly countries of the south—than it is here. That is to say, it is the mathematics department that does theoretical mathematics and statistics, and then the so-called statistics is done by people in all kinds of fields with very little statistical training. There generally is very little thought given to the use of statistics in public policy. In talking about international perspectives, there needs to be more attention given to what we are doing about training people to work internationally.

S. RAO JAMMALAMADAKA: The split between the theoretical and applied statistics exists everywhere, to varying degrees. As I noted, in some places such as Sweden, it is even formalized with separate departments for mathematical statistics and statistics. In many countries, such as Japan and Australia (for instance, at the Australian National University in Canberra, where I taught), the statistics department is located in the Faculty of Economics and Commerce. In my own university, although we have a department of statistics, there are people in psychology and sociology, who are not trained in statistics, who teach their own statistics courses. This also happens at the University of California at Berkeley, as it does on many U.S. campuses. I do not think it is necessarily bad. There are some courses with large substantive content that the faculty in those other departments may be able to handle better—as long as they know when to send their students in our direction. For instance, in Sweden, the industrial engineering department sent their engineers to the mathematics department for a course that I taught on reliability, quality control, Taguchi methods, and so on. Based on a limited sample, my general conclusion is that there is no detectable "international" (as opposed to the "intranational") trend regarding the theory-versus-applied split.

CHAND MIDHA: I have heard many comments from people in statistics departments. Ours is a department of mathematical sciences that offers a master's degree, and there are many similar departments. The problem is that, although we have offered some good service courses, many of the other university departments are trying to offer different courses, some of which are being taught by so-called statisticians. We get little chance to develop any interests for these students so that they might come and consider going further in statistics. Since yesterday's session, I have had the feeling that more students should have more exposure to other things, should have double majors, and so on. I feel that if we attach some importance to these required courses and people who have other things wind up taking minors in other subjects, we may lose many of our statistics students to other areas.

JOHN TUCKER: That sort of problem is not particular to statistics. A report in the *Notices of the American Mathematical Society* (April 1990, pp. 408-411) pointed out that in this nation more of the advanced undergraduate mathematics instruction is being done in non-mathematics departments than is being done in mathematics departments, in terms of the total number of students taking those kinds of courses.

FIENBERG: I used to say things similar to what Chand Midha's comment suggests, but over time I have taken a somewhat different attitude and perhaps have provoked people as a result.

You have to be careful when you allow statistics courses to be taught in other departments. Yet often they are taught by people whose qualifications to do real statistics may be superior to those of people in the mathematics department or even in the statistics department, and the challenge that statisticians face on campus is how to draw those people into interaction with the students whom we want to learn statistics in a way that is not in competition with the elementary course. Rather, a distinction needs to be made between those people who are doing quality statistical analysis wherever they are based, and those people who are perhaps not qualified to do that kind of statistics instruction.

EDDY: I want to amiably disagree with my colleague. In my limited experience, the statistics courses given by other departments tend to use statistics as a collection of procedures and tools for, as someone said yesterday, feeding on particular problems: "If all you have is a hammer, everything looks like a nail." So I disagree rather vehemently with Steve Fienberg's perspective.

SNELL: I want to amiably disagree rather vehemently with Bill Eddy's view. Certain things in Dartmouth are exactly what Chand Midha is describing. The fact is that social scientists and economists long ago learned to analyze data that arises in their discipline. Having thought about the problem of how well statistics is taught in other departments, I believe they have for years been using exactly the kind of techniques everybody has been advocating, namely, motivating the theory from the application setting.

BAILAR: Perhaps we can all agree that these people in what we might think of as the hinterlands, the outlying departments, are at least *perceived* as offering better training than that which people who get into the statistics department are perceived as obtaining. And maybe the first task is to deal with that perception, and so to produce a greater tendency in students toward taking the courses in the statistics departments that have the solid core of understanding we believe underlies this whole enterprise.

KETTENRING: I do not really care where they are trained. What I care about is what they can do when they are on the job. In a larger sense, we really have not done a very good job in discussing the need for radical changes, and so how are we going to know if we are succeeding when we start making needed changes? We need to agree upon or have a dialogue about some measures of success or failure, because I think we can get blocked out here discussing who can teach what.

I have some criteria from my industrial perspective that I would use to judge what is being produced. I would look at the titles of the PhD theses of the statistics students who graduated from your departments in the last year and a half, and ask, "Why do this research? What is its impact?" I might even try to read some of them to see if I could understand them. I have my own private ways of trying to assess such things, and in the context of these discussions of changes, I believe we need to develop criteria for monitoring how the statistical community is doing in making needed changes, so as to avoid just talking.

RONALD THISTED: As to other departments offering introductory courses or even advanced courses in statistics, at the University of Chicago, we do assess the effectiveness of the statistics program.

In our department, we see other departments offering statistics courses as partly a problem and partly an opportunity. We believe that if other departments' students shy away from our courses, then perhaps we are doing something wrong and perhaps their courses are meeting a need that our courses do not. So we try to make it our business to find out what needs those courses are meeting, and if we think we can do it better — and usually we do — then we try to do two things. First, we try to change our courses to do a better job of meeting those needs, and second, we try to do a better job of communicating to our colleagues in other departments just what it is we really do as opposed to what they think we do in the courses we teach.

GRAY: Going back to the original question that started this line of discussion, if you have people teaching statistics courses in other departments, one practical suggestion is to invite them to teach some sections of the courses in your department. American University has a regulation that all the elementary statistics for undergraduates must be taught in the mathematics department, but people from psychology or economics are brought in to teach sections of the course because there are many, many sections of it. This approach also has the advantage of providing some opportunities for people to do collaborative work because they have to talk about what they are teaching.

We convinced the university administration to let this be done simply as a more economical way to do it, because if all the sections are under the control of one department, enrollments can be better managed to ensure fully enrolled courses that cost less money. Given that most universities are now worrying about costs, that is a very good reason for getting started. But I think an equally important benefit is that it is much more productive, especially in bringing in people who are doing substantial statistics, and who in many cases do have very good training and very good ideas, to work with the statistics department.

BAILAR: There are a couple of things that I am not hearing about here that might in time become pretty important. One is to explicitly acknowledge that statisticians cannot be all things to all people, and statistics graduates cannot be all things to all people. There is a need for specialization. Some of that is already recognized, as exemplified by biostatistics and by the fact that potential statisticians do spin off to other departments. But perhaps more explicit thought should be given to what kind of specialization in graduate training programs would be appropriate.

The second thing I am not hearing is comments of the form, "How can I go back and make this change?" I have the feeling that many of us think, in effect, "Gee, things are going pretty well now; it is everybody else that needs to change." This leads me to suggest, although with hesitation because I am not a CATS member, that — if they have not already planned to do this—perhaps CATS would develop a panel on model curricula embodying the changes discussed here that people could examine and adapt for their own use.

JOHN TUCKER: In fact, this symposium is viewed as a first effort in trying to instigate change, and producing model curricula is considered one thing that sometime has to be done as an ultimate outcome. And CATS is indeed an appropriate body to consider it.

LAWRENCE BROWN: This issue of what faculty with what background and attitudes are appropriate to teach statistics raises the question, Is statistics different from any other discipline in that respect? I do not believe so; I do not believe that social scientists who are teaching statistics courses have any different attitudes toward service courses. That is not a distinguishing characteristic.

BAILAR: Let us see if statisticians can be better than these others.

TUCKER: The observation of having years ago given up some rights to teaching these service courses in statistics is, again, not confined to statistics; mathematics has had the same thing happen, and it is also a matter of concern to mathematicians.

JOE WARD: I have been teaching at the University of Texas at San Antonio, after having taught at Clemson, and am spending a lot of time on what I believe disciplinary or institutional inertia is impeding. When we consider what a model curriculum would be, we must keep in mind where we are going, what our objectives are, how we are going to get there, and how we are going to know when we are there—that is, assessment. Those things need to be addressed, because if you decide to have more probability and statistics but continue teaching it the same old way and do not change to a way of teaching that incorporates computers and uses them routinely, substantive change in the results is unlikely. Unfortunately, people are generally still teaching the precomputer approaches to doing statistics. A new edition of a book may be 200 pages longer than the old one, but the book has the same form, style, and approach as the one somebody saw earlier because that is what sells. The objectives need to be looked at, and what is to be done with contemporary, real-world problems in technology must be decided. I call it the talk-down approach. Start with the problems, and then look down the microscope to see what the key issues are that you are trying to teach. This approach, I think, would eliminate a lot of the less crucial material.

ROTHMAN: Concerning the issue of teaching correct theory, all our theories are correct in some world. The problem we have to face is usefulness, predicting the particular situation. I do not believe the statisticians are closer to that truth than some other people are, whatever their department may be. In fact, a lot of what statisticians do is simply not useful. There are confidence intervals that allegedly work 95% of the time in the abstract world and work only 50% of the time in this world. Regarding the issue of bias, there is no question that it is a substantial fact of life with which we must deal. Statisticians are not closer to a solution than their colleagues in other fields. Statisticians need to pay attention to what those colleagues have to say.

DONALD MYERS: Stemming from what we heard on cooperative learning yesterday, should we begin to talk about cooperative rewards? To some extent, what happens to the statistics department, the mathematics department, or any other academic department is based on the reward system. The particular direction that statistics takes, the courses that are offered, and which are emphasized is largely a matter of how we are rewarded by the university. At present that reward tends to be bestowed on the basis of individual activity. Teaching, generally speaking, is not an individual activity; how successful you are depends on how successful your predecessor was -- that is, the person who taught the students before they got to your course -- and you in turn affect the next instructor. The reward system does not recognize that, nor does the way in which research money is given out recognize that.

A second point has to do with remarks that Phillip Ross made yesterday on the problem of a manager facing a decision that had to be made on the basis of imperfect information. The manager does not happen to be a statistician. He would give the statistical analysts two months to do a statistical analysis on the problem. The manager in a sense did not care what the statistics looked like. But a decision had to be made, and the decision was selecting from two different possibilities, one of which involved a $40 million investment and the other a $60 million investment. In one sense they had a relatively small amount of data, yet in another sense they had a relatively large amount of data. They also had something called soft data, namely, the expert knowledge of certain geologists, and they had to incorporate that initial information in the statistical analysis. That is a perspective I do not think I have heard in any of the comments about how to incorporate, for lack of a better word, the expert knowledge of people into the statistical analysis.

TUCKER: That situation is addressed in the 1992 CATS report entitled *Combining Information* that was originally published by National Academy Press, and that the American Statistical Association has just republished as Volume 1 of ASA's new Contemporary Statistics series. The report also addresses issues that arise in settings such as the one that confronts EMAP, as described yesterday by Phillip Ross.

A University Statistics Program Based on Quality Principles

Edward D. Rothman
University of Michigan

Introduction

The purpose of this paper is to present aspects of a university statistics program that have had a positive impact on the student's joy in learning. The program's plan is to teach students an efficient learning process aligned with the purpose of a liberal arts education. Implementing the proposed program nationally will require changes in curriculum, research, and the pedagogy, and as a result will require faculty to adapt. When students become more efficient learners they increase their joy in learning.

Efficient learning depends especially on asking appropriate questions. The statistician can then provide suggestions regarding the purpose of the study, the measurement process, study design, the planned analysis, and the presentation of results. To be useful we must understand the limitations of statistics and then make clear what can be said or when there is a leap of faith. This notion of being useful is substantially more demanding than correctness or appropriateness. We want to communicate information about the client's world rather than the idealized world that never obtains. This is the basis of the Deming message about differences in analytic and enumerative studies (see, for instance, Deming, 1993, 1990).

Consider that much of what statisticians teach is based on samples drawn from a single urn so that we can say something about the entire urn's composition. The reality is that we take samples from a succession of urns and we hope to say something about an urn that has yet to be sampled. Is it a wonder that 95%-level-of-confidence intervals do not cover the parameter of interest as often as expected? The systematic biases are important but the way we put it all together is critical, and neither are given adequate attention in our programs.

System thinking is much easier to describe than to implement. One feature is that the performance of the system is viewed in terms of impact on the "bottom line," and so interaction is a key. The effectiveness of measurement process, the design, analysis, presentation, and every step need to be considered together. Optimal designs and decision rules for conditions that do not obtain are not useful. Our clients are better served when we recognize this problem and teach this coordinated approach in our classes. It is far more than the study of robust design or analysis. We need to pay attention to the purpose of the study and every element involved in the study.

The proposed program must also take account of variation. Especially evident are the varying interests and talents of the students and faculty. Whatever program we arrive at must be flexible. Much of what we see is a deterministic take-it-or-leave-it solution. Consider how little flexibility exists within and between our programs.

In the language of "quality," the variation is reflected in two distributions. One, labeled the voice of the customer, reflects the distribution of customer preferences while the other, called the voice of the process, represents what the process delivers at a point in time.

Achieving the aim means making the distributions congruent. And to achieve congruence we require a program that is central to a liberal arts education.

At the University of Michigan, alignment of all programs with a liberal arts education is supported by our administration. Consider that in every promotion case and in reviews of our department, we have been asked to describe how the area or subject is central. The response by departments to the administrative challenge has been slow. Few programs teach system thinking, place substantial emphasis on measurement, distinguish between analytic and enumerative studies, or are very concerned about graphics and presentation. The pedagogy must also be altered if we expect to establish a pattern of lifelong learning among a broad cross section of students.

Voice of the Customer

Although employers, parents, faculty, and others have a substantial claim as the primary external customer, we at Michigan look to our students first as indicators of the success of the proposal. Their increased joy in learning could be measured, in part, by enrollments, but other indicators may be considered.

The voice of the customer not only varies between customers but can also, as with any service or product, be affected by others. The business community might call this marketing or advertising, but we will not be quite so crass. Contact with students at an early stage to discuss what they ought to be concerned about is important. Though our preference would be to have a faculty member in the counseling office on a regular basis, we do not have the resources. Instead, we talk to some students at orientation, and the message we share with them manages to make the rounds of the counselors.

These orientation talks have met with substantial success. Though they are only thirty-five minutes long, we include a statement about learning, read some poetry, and present a couple of examples of sense deception. The poem by Yeats, "For Anne Gregory," is used, but there are many alternatives that reflect the difficulty of "seeing" or learning by observation. Sense deception examples are abundant, too. We select examples based on aggregation and coincidence.

Among the students addressed, about 20% elect an introductory class in statistics during their freshman year. The message that statistics is central to a liberal arts education and that they are here to learn how to learn rather than simply accumulate facts, which may soon be seen as incorrect, is quite compelling.

Beyond this introductory role, it is important to understand who our students are and what they need and want. Students want to be productive. They want to contribute to society and also be viewed as individuals. Though these two desires are sometimes in conflict, the balance between the two needs to be addressed at the individual level. We are not, however, and should not be in the training business. Industry must expect to be lead in this activity. What we need to provide, as the Peace Core message suggests, are lessons in fishing and not simply fish dinners.

Once we identify learning styles and student backgrounds, alternative options can be provided. Some students may benefit through early involvement in research. Our summer research opportunity program for minority students here has had some success. Other students benefit from team learning experiences. These allow the teaching staff to spend time with individuals while most students are engaged in the group activity.

Voice of the Process

The education process is dependent on the faculty, administration, equipment, and materials. But the faculty is our most important challenge. The faculty may be unwilling to change, may have had little exposure to pedagogy, and many may not have worked as consultants in applied research.

We expect coaching will improve the quality of the teaching in a very substantial way for those faculty interested in improvement. (Somehow writing these words now is easier than when I was chair of the department.) However, changing what is taught is another matter. There is a general discomfort with anything new, especially if we require radical changes, in the curriculum. The proposed countermeasure is to broaden our focus with new faculty, continuing education of present faculty, and by allowing those who are more comfortable with theory to broaden their view of the theory.

The applied statistician would need to become much more involved with the integration of all elements required for efficient learning. Theoretical statisticians would continue to develop theory but would broaden the focus by consideration of more general objectives. It is especially important that they become more involved with the metaphor of the theory.

For example, they might ask what the theory says about what is the appropriate question. In Robert Pirsig's *Zen and the Art of Motorcycle Maintenance*, he asks, "If the cycle doesn't start should we ask what is wrong with the starter or why doesn't the cycle start?" Deming as part of his four-day seminar illustrates the issue through the "Red Bead Game." The game is a physical simulation of a sequence of identically distributed random variables. The random variable is the number of red beads selected from a bead basket containing several thousand red and white beads. The paddle has 50 holes and is filled after much fanfare from the basket. Participants, "the six willing workers," take turns drawing samples from the basket. There is much to learn from the game, but the relevant issue is that it is not useful to ask why a worker has a particularly large number of red beads, because the answer does not depend on the worker at all.

We need to engage in continuing education of our faculty to sustain the improvement. Some faculty might benefit from sabbaticals that allow them to study with non-statisticians, and the forced reductions in summer moneys available to researchers may encourage some to spend part of their research time engaged in consultation. Our consulting laboratory, the Center for Statistical Consultation and Research, encourages faculty to provide several hours per week in consultation with other researchers. Graduate students employees of the center screen requests for consulting time so that it is more likely that the faculty time is productive. The result is often a collaborative research effort that yields a higher-quality product.

The departmental seminar series can be modified to broaden faculty focus. Seminar speakers from non-statistical disciplines can be primed to initiate a dialogue that can lead to learning. The dialogue would generally require someone to keep the seminar moving forward, and there may need to be some discussion of the purpose of the dialogue.

The provision of reference material on the use of theoretical statistics in the broader context of learning would help, too. However, the market for a radically new product is less likely to be published, and we need a mechanism to encourage such efforts.

Curriculum Principles

The curriculum described below focuses on what we should teach but does not say how. We expect that the topics selected are generally consistent with the backgrounds of the current faculty, but changes in the faculty are essential. Use of external resources including faculty in other fields, adjunct professors, video, and other sources of material would also enable a more enriched experience, and team teaching may also prove effective.

The general principles advocated for the curriculum include the following:

1. Courses should have material aligned for a common purpose.

As proponents of efficient learning, we need to bring coherence to the program. Each course should have a clear, stated purpose consistent with the overall aim, and context setting is especially important. Students benefit from a discussion that indicates how the course work fits with the overall aim.

There are many ways to achieve a common purpose, and we are not advocating a single curriculum for every department. Indeed, the needs of the customers will be better served with a richer variety of options.

2. Subject matter should be integrated into the entire curriculum.

Applications of what we learn requires integration of ideas and skills. Our models and approaches to learning have become too compartmentalized. Remember, we want to enable students to apply their knowledge to situations beyond the limits of the course. An appreciation for some model requires that we know both when it is useful and contexts where it is not.

Rather than having courses devoted solely to large sample theory, for example, and having the large sample theory covered only in such classes, we propose that the material be integrated into virtually every class. History of statistics classes, too, should also be integrated. The ideas should provide a foundation for the questions we ask and should establish a more scholarly approach to learning.

3. Subject matter should be useful at several levels.

Material should yield benefits at several levels, as some literature does. Lewis Carroll's *Alice in Wonderland*, for example, can be studied and restudied. Our students, and all students, should have the opportunity to harvest what they are capable of handling. Using material that can be revisited allows for efficiencies and accommodates variation between students.

The introductory class, Statistics and the Scientific Inquiry, involves lectures by a faculty person and a recitation section led by graduate students. The comments from the graduate students are that they learned as much from this course as they did in any other class. It is likely that the perspectives they bring to the material designed for freshman allowed them to learn at a different level.

4. **The curriculum should be more heavily weighted toward measurement and more generally toward design of the study rather than analysis.**

Efficient learning requires a well-designed study. This means careful attention to the purpose of the study, the measurement process, and to the general design. Our courses need to reflect the fact that most studies are observational rather than randomized.

The current curriculum is too focused on analysis, and the measurement process is virtually always taken as given. Discussions of variation are based on analysis issues, and even though some design has been included in our introductory courses, much is still devoted to aspects of the analysis. Further, the analysis has been concerned with questions of statistical significance or the sampling variation associated with a certain repeatable process. Neither question is as important as we suggest, and, further, there are many more important questions that are not asked.

5. **Models and other characterizations of results should attach more weight to physically based models and not be simply mathematical expressions of the analysis.**

Our inclination is to seek mathematically simple models. Sometimes these models are not simple when seen through the eyes of a scientist. Physically meaningful models should be given at least as much attention as those that are mathematically based. The parameters should have physical meaning and, to the extent possible, describe (uncorrelated) features of the data.

6. **Statistical theory has much broader implications than as a framework for data analysis. These broader implications should be built into the curriculum.**

The major emphasis here is on the provision of questions. A Talmudic scholar once said that a good question is more than half the answer. Our theoretical models need to provide investigators with questions.

Some Suggestions for the Curriculum

The following sections provide some detail on what we should include in courses. This is not proposed as a comprehensive list but as a means of encouraging a different perspective from that we acquired when we were graduate students. The intention is to describe subjects other than data analysis that lack attention.

Measurement

Although we complain about the requests to perform analyses on poorly designed studies, we sometimes take for granted the quality of the measurement process, and this process of measurement is clearly an integral part of the design. Much of the material in our statistics texts and research papers begins with the measurement process as given, yet we have much to share about this process.

A measurement is defined by answering three questions: Why are we making the measurement?, What is the measurement?, and, How is it measured? The "why" question needs to be addressed because we want our system approach to learning to be aligned with a purpose.

The statistician has much to say about what is measured. Does the measure contribute to the purpose? How many dimensions are required to characterize the quantity? Measurement always causes us to focus on an aspect of the purpose and is thus an abstraction or model. The "what" needs to be returned to throughout the curriculum. At one level, we could discuss scale and units of measurement, and at another the dimension question can be raised.

Finally, we need to define the method used to measure. Each method will yield differences, and comparison is not possible without a common operational definition. Systematic changes arising from different methods of measurement can clearly be far greater than the sampling variation. And for a particular method of measurement we need to consider systematic biases. The discussion should not be relegated to a course in survey research but should be addressed throughout the curriculum.

Variation

Conventional courses discuss the measurement of variation in an abbreviated manner, place much attention on the standard deviation, and then describe the connection with probability through Chebyshev's inequality. In addition to these topics, the new curriculum should also emphasize the selection of an appropriate measure of variation and should describe the factors that influence the measure.

The selection of an appropriate measure of variation is like the choice of any measure. We find that students understand the measure if they are involved in establishing the purpose. Should the measure reflect the number of items studied or how different they are? If we add an additional source of variation, then should our measure reflect that fact? What units should the measure possess? Should the measure be simple to calculate and have a straightforward interpretation? Each question allows us to say something useful about standard measures such as the range and mean absolute deviation.

Use of the standard deviation with angular measures might provide a poor reflection of spread. The dependence on the choice of origin is clear. Variation in nominal data can be measured in many ways: an entropy measure or the homozygosity measure in genetics are just two examples. The questions raised in the preceding paragraph have broad implications, and students will benefit from the general discussion.

Variation is, in general, increased with aggregation, differences in operational definitions, complexity, and over-control. The current curricula do pay substantial attention to effects of aggregation, but the other three generic sources of variation are given little attention. Demonstration of the effects of over-control is facilitated through the use of the Nelson funnel experiment.

In the experiment, beans are dispatched through a kitchen funnel placed above a "target." The funnel is held, initially, in place about 8 to 10 inches above the target. After a bean is dispatched and the point where the bean comes to rest is recorded, the funnel is either left in place, rule 1, or its location adjusted according to the outcome. The objective is to minimize the average of the distances between the rest positions of the sequence of beans and the target.

There are various adjustments. Rule 2 moves the funnel from where it last was by an amount needed to compensate for the last error. Rule 3 also examines the last outcome, but compensates from the target rather than the last funnel location. Rule 4 requires that the funnel be placed above the last point where a bean came to rest.

Rule 1 leaves the funnel in place above the target. Though it is clearly the optimal strategy, it is analogous to management sitting on its hands and is not what many people want to do. Rules 2 and 3 increase the average distance, but rule 4 is most commonly used. Rule 4 produces a sequence of distances that we can model as a random walk. The implications on the variation are clear to a statistician. What students find interesting is that the failure of many common practices can be seen through this process. Making the next house key from the last one and the children's game "telephone" are two examples. The purpose of these analogies, and of similar ones for rules 2 and 3, is to understand the impact of over-control on a stable (that is, stationary) process.

Learning how to reduce the variation in the funnel experiment is also generally informative. Three systematic changes are lowering the funnel, using a sticky material around the target, and cooking the beans. Each of these actions is not a response to the last outcome but does change the "common cause" of the variation.

Mathematical Statistics and Probability

The theory of statistics and probability will continue to play an important role in the curriculum. Teaching students about deduction, asymptotics, and abstraction is helpful in the construction of useful theory. However, the separation of theory and applied statistics is counterproductive. Our overall purpose requires that we diminish the barriers between areas and approaches to knowledge.

Asymptotic theory provides a case in point. Some students do not appreciate the importance of large sample theory to the modeling effort. The division also creates a potentially dangerous view of what we do, especially in the eyes of an administration that considers application an inferior investment.

The current curriculum in most statistics departments is, however, entirely too focused on hypothesis testing. We are not providing our customers with useful information when we do this, especially because we neglect far more important models. Like the theory of best linear unbiased estimation, the hypothesis testing theory is straightforward, fun, and easy to teach, but the view that we are sampling from a basket of balls does not provide a useful model for most applications. Indeed, it is much better for us to imagine that each observation comes from an entirely new basket and that our purpose is to learn about a basket that has not been observed. The "leap of faith" needed to make such inferences should be made explicit.

Analogy and Metaphor

The curriculum should also make more substantial use of analogy and metaphor to generate working models. The current process moves from assumptions and a deductive argument or from data to density estimation. We propose that analogy be used to generate models, too. Cognitive psychologists suggest that such learning is quite common.

Examples are plentiful, especially in portions of the curriculum dealing with deterministic systems. Newton's law, $F = ma$, connecting an action, F, with a reaction, a, through a proportionality constant, m, has analogies everywhere. Ohm's law, $E = RI$, and *blood pressure* $= R \times$ *cardiac output* are two examples. The force is a voltage or pressure on the left side, and the reaction is current or cardiac output on the right, and in these two cases the proportionality constant is resistance. Indeed, it is difficult to find a field without a first-order kinetic law.

What corresponds to such cases in statistics is not clear. There are abundant urn models, but many of the more exciting ones may not respond to simple techniques (that is, techniques that do not require some mathematical maturity). When we oversimplify, much of the flavor is lost. However, computer simulation to illustrate the behavior of solutions can be much more instructive than an analytic approach that can cause some students to lose the forest for the trees.

Asking students to suggest analogues of theorems is a useful learning device. It helps them appreciate the conditions and the implications of the result. In a freshman class, we at the University of Michigan have students describe situations in which a random walk might describe a process. They have little problem with the exercise, and the usefulness of the result is much clearer to them. For example, they have a deeper appreciation for the growth in the standard deviation when they ask what happens to a message that is passed from one person to another with some noise added each time.

Graduate students in statistics may know Wald's identity but may not understand what it says as clearly as when they apply it to situations outside a sequential analysis class. Several years ago, a biologist colleague of mine, Julian Adams, returned from a AAAS conference that included a talk about preferences for male offspring and the impact on the ratio of sexes. In the talk, it was suggested that such preferences by Roman families increased the proportion of males in their society. Wald's identity is not common knowledge in our introductory classes, but even in graduate classes, students may have a difficult time making the transition from theory to use.

In an enlightening paper by Tversky and Kahneman (1974; see also Kahneman et al., 1982), we learn about other difficult transitions. These involve the regression effect and the law of large numbers. Even after studying the law of large numbers, many students may not know that a hospital with 50 births per day is more likely to indicate a wide disparity between the number of boys and girls born than is a hospital with fewer births. Even after a class on the regression effect, students may find a more complicated explanation for the subsequent success of a baseball player's use of a rabbit's foot after a poor start.

Model Description and Presentation

The representation of models in algebraic terms also limits their usefulness. Geometrical and algorithmic expressions provide a richer symbolism, but much remains to be done in providing vehicles and strategies that convey understanding. Although statisticians have some interesting thoughts and excellent examples involving the presentation of a complicated story in simple terms, we need to develop a more general theory or set of principles for the presentation of results. Although much has already been accomplished (see Cleveland, 1993a,b; and Tufte, 1983, 1990), more progress in this direction is needed.

Bias

Compared to variation, systematic bias is given little attention in much of the curriculum, with the exception of the sample survey class. But this lack of attention is not reflected in the applications we encounter. Indeed, we act as though randomization, blinding, and other experimental techniques eliminate bias, and we know that they do not do so. In the introductory classes it is important to identify the various forms of systematic bias and when possible to provide a quantitative analysis of the impact of these biases.

Selection biases, including size- and length-biased selection, are particularly important. The direct effects can be developed easily, while the indirect effects can be introduced for identification purposes only. For example, if we want to learn about the average area of farms from a sample of farms selected by choosing farms hit by a dart thrown at a map, we can see that large farms are more likely to be included than smaller farms. Similarly, the indirect effect situation is characterized by a screening program in which individuals with longer latency periods for the disease are more likely to be selected than those with a shorter latency period.

The impact of a low sampling rate is described in upper-level time-series classes, but the effect of aliasing is of general importance. Simple versions of it can be presented in introductory classes to indicate the impact, and then a mathematical description can be given in the advanced classes.

Generally, we act as though the samples we select are drawn from the "basket" that represents the target for our inference. This notion has little foundation and causes others to see statisticians as less useful. The textbooks are filled with calculations based on this premise yet have little helpful information to assist us in the "leap of faith."

Sense Deception

Even when students have been exposed to a wide variety of statistical concepts, they may not be able to identify everyday situations in which these concepts are applicable. Tversky and Kahneman provide many examples of this with special attention to the law of large numbers and the regression effect (see Tversky and Kahneman, 1974; Kahneman et al., 1982). Also important are aggregation and coincidence.

There are many examples of the aggregation effect in our standard textbooks. Students find it difficult to appreciate the impact of a third variable on the relationship between two variables studied. Since the third variable may not be unknown, conclusions regarding the relationship between the other two could under certain circumstances be reversed. It is curious that in spite of these examples, desegregation techniques are not more prevalent in our courses. Of particular importance is a paper by Joiner (1981), who provides several examples that offer clues to the existence of lurking variables and how conclusions need to be reconsidered.

Design of Observational and Randomized Experiments

Much of the data we study arises from observational rather than randomized studies. Such studies come in retrospective and prospective varieties, and our students need to know how to identify the type of study and decide what are appropriate questions to raise.

We also need to reexamine the purpose of our studies. Much attention is placed on attempting to learn about differences that we know exist anyway. There are many other purposes that require different approaches. Understanding what a relatively flat or sharp response surface entails is one example.

Pedagogy

Grades

Students may not be focused on everything we talk about in class, but they certainly pay attention to the bottom line, the final grade. They want to know how they will be graded and what is required to achieve a high grade. However, we know that there is a conflict between learning and preparation to obtain a high grade. Test preparation activities encourage short-term memory, cause students to compete rather than cooperate, cause them to focus on doing the minimum required to achieve a target, and create general unhappiness with learning.

The unhappiness resulting from the grading process stems, in part, from the use of grades as an incentive for an activity that is intrinsic. Learning is a desire that virtually all are born with. Incentives such as pay for performance create an alternative purpose that can undermine the intrinsic desire for the activity of learning. Even students who receive a high grade may not be happy if they feel that the reward is overjustified.

The grade incentive does, however, facilitate classroom management. Students become quite attentive when they hear, "The test will include. . . ." But the short-term advantage of grades as a means to get students to focus in class undermines the long-term goal of continual learning.

Grades do provide feedback. However, feedback is more effective when it leads to enhanced learning; grades tell us something about where we were and not how to get ahead effectively. Alternatives to grades are provided below.

Mastery

Conventional tests, mid-terms and finals, are often inspections of what has or has not been learned. The usefulness of the results is limited, and the final examination comes too late to provide students with help on the current material. It is seen as a statement of closure rather than as a learning opportunity.

There are several alternatives to these tests. Only one alternative — the evaluation of mastery -- is described here. The approach is similar in spirit to the Keller "Personalized System of Instruction" program (see Keller, 1968). We at the University of Michigan continue to experiment with variations on the implementation of this evaluation in our class, Statistics and the Scientific Investigation.

In this course, there is a natural division of material into ten fundamental principles. Demonstration of mastery requires that students first complete prerequisites and then demonstrate that they have the appropriate facts, understand the principle, can apply the principle to a situation that is unique, and can "explain" each of these steps in a clear fashion to an "educated layperson." The results are evaluated as "mastered" or "not yet." Masteries are not failed, and when students receive a "not yet" designation they can retake versions of a particular principle.

The prerequisite serves to place some of the learning responsibilities on the students' shoulders. However, students sometimes choose to take a mastery evaluation that has a more narrowed focus of what was wanted than the lecture provided. This is wasteful of teaching fellows' energy and undermines learning.

Initially, a homework assignment on related material served as a prerequisite. Students needed to complete this ungraded assignment before they were allowed to take the mastery evaluation. However, feedback on the homework is needed to focus attention on students' weaknesses. Software is being developed to provide on-line tutorial help with each principle.

Journal

If the purpose is to teach students to apply what they have learned to entirely new situations, then the evaluation ought to include such cases. University of Michigan students apply what they have learned to a collection of newspaper, magazine, and scientific articles of their choice. Each page of their student journal focuses on a principle.

Since we want the activity to be exciting, students choose the articles, but we prescribe the format. Each journal entry identifies a principle, asks a question based on the principle, and then provides an explanation as to why the question could lead to a different perspective.

For example, a recent headline in a local paper indicated that whereas 40% of black applicants had applications for property insurance rejected, only 10% of non-black applicants had their applications rejected. A student asked whether family income was a factor. He argued that it might be that both for poor applicants and for non-poor applicants, rejection rates were higher for non-black applicants, but appeared lower when aggregated. The student then indicated with a numerical example how this could happen.

Although we have had some problems, especially with some dishonest students, the results of this effort are the most exciting by-product of the changes made.

Feedback to Instructors

The purpose of feedback on students' understanding is to provide both the student and the instructor with direction. The feedback should be frequent, it should provide an opportunity to improve, and it should focus on a demonstration of learning and not on the accumulation of facts. Student evaluation of teaching at term's end is useful but is too little and too late.

The one-minute paper proposed by others can be adapted to improve teaching. By providing comprehensive timely feedback, it causes students to focus on the lecture, and it helps instructors to provide a firm foundation.

We ask students two questions during the last minute or two of class. First they identify questions that they believe they can now answer as a result of the material covered, and then they identify questions that they would like us to answer to help them understand the material at a deeper level. Feedback based on their responses is provided at the beginning of the next lecture.

Feedback to students must be timely and organized. A Pareto chart that displays the frequencies of responses from most to least frequent can be used to summarize the information. The few most frequent response categories can be reviewed at the start of each class. Subsequent revisitings of a topic can be made more efficient by orienting the discussion in a manner that anticipates the problems.

Discovery

Focusing on the lecturer is not a long-term solution. What we need to provide are fishing techniques rather than the seemingly more expedient serving of fish. One approach is through the process of discovery. Although discovery is seen as a slower process for learning prescribed material in the short run, it can facilitate learning when lectures end.

Courses on design of surveys and experiments, as well as probability classes, are prime candidates for discovery. In the survey design class and experimental design classes at the University of Michigan, students design surveys and experiments and carry out the design. Obtaining first-hand information on what could go wrong is more effective than listing potential sources of error. In analysis too, the appreciation of interactions is enhanced when the interactions are discovered. Probability classes, when coupled with a computer laboratory, provide an excellent opportunity for discovery. Appreciation for scaling, in particular, is greatly enhanced when students experiment with the choice of scales.

Group Learning

Learning is enhanced when students can share. Expressing what they think they understand to others and hearing what others can contribute are both worthwhile activities. Each requires some interaction with an expert, but the interaction should be limited.

Summary

University classes should emphasize the process of learning rather than the presentation of facts. We need to view what we do as educators as distinct from training, which should be done on the job. Students want to learn something that they view as useful both today and tomorrow, and the resulting tension should be seen as an opportunity to improve what we do in our statistics programs.

We should view education in statistics as part of the system of education. It is essential that what and how we teach, and the connections or interactions between statistics and the other parts of the system, be aligned for this common purpose. To view statistics education as separate from university education is the antithesis of the proposal.

This paper has focused only on aspects of this transformation, much remains to be completed. The proposed transformation involves changes in some courses, a major emphasis on design of studies, and a substantial focus on critical reasoning. And the way we teach needs to change, too, from a lecture format to discovery, group learning, and a general recognition of variation among our students and faculty. Finally, we need an honest examination of the way we evaluate students and a replacement for grades.

Although enrollments in statistics classes at the University of Michigan continue to grow, the growth is not broad based, and few of the students choose graduate school in statistics. Standard classes have remained stable or have declined, while courses that focus on the role of statistics in learning grow at an incredible rate. Statistics and the Scientific Investigation grew from 80 students per semester to over 250 in four terms.

References

Cleveland, W. 1993a. *The Elements of Graphing Data*. 2nd ed. New York: Van Nostrand Reinhold.

Cleveland, W. 1993b. *Visualizing Data*. Summit, N.J.: Hobart Press.

Deming, W. E. 1990. *Sample Design in Business Research*. New York: Wiley.

Deming, W. E. 1993. *Quotations of Dr. Deming: The Little Blue Book*. 2nd ed. S. McCrea, ed. Fort Lauderdale, Fla.: South Florida Electric Auto Assn. 128 pp.

Gnanadesikan, R., and J. R. Kettenring. 1988. Statistics teachers need experience with data. *Coll. Math. J.* **19**:12-14.

Joiner, B. 1981. Lurking variables: Some examples. *Am. Stat.* **35**:227-233.

Kahneman, D., P. Slovic, and A. Tversky, eds. 1982. *Judgment Under Uncertainty: Heuristics and Biases*. New York: Cambridge University Press.

Keller, F. S. 1968. Goodbye, teacher. . . . *J. Appl. Behav. Anal.* **1**:79-89.

Tufte, E. R. 1983. *The Visual Display of Quantitative Information*. Cheshire, Conn.: Graphics Press. 197 pp.

Tufte, E. R. 1990. *Envisioning Information*. Cheshire, Conn.: Graphics Press. 128 pp.

Tversky, A., and D. Kahneman. 1974. Judgements under uncertainty: heuristics and biases. *Science* **185**:1124-1131.

Velleman, P. F., and L. Wilkinson. 1993. Nominal, ordinal, interval and ratio typologies are misleading. *Am. Stat.* **47**:65-72.

Chance: A Course About Current *Chance* News

J. Laurie Snell
Dartmouth College

I was asked to talk about some of the successes and failures that we have had at Dartmouth College trying to make students in statistics more aware of the interdisciplinary nature of their subject. Naturally, I would rather speak about our successes than our failures.

We have had two years' experience with colleagues at other colleges developing and teaching a new introductory course called Chance. This project has been supported by the National Science Foundation's Undergraduate Development Program and the New England Consortium for Undergraduate Science Education.

The Chance course has been taught or is being taught at Dartmouth, Middlebury, Grinnell, Princeton, Spelman, and the University of California at San Diego. While it has not been taught at the level that we have been talking about in this symposium, it could be, and certainly some of the approaches we use would enrich more advanced undergraduate and graduate statistics courses.

Chance deals with current events that rely on statistical and probabilistic concepts. We got the idea for the Chance course from reading *Chance* magazine. This magazine was started by Springer-Verlag in 1988 and now is a joint publication of Springer and the American Statistical Association. One of its founders, Stephen Fienberg, is participating in this symposium.

In its brief existence, *Chance* has attracted some of the leading workers in probability and statistics to write articles that present important statistical problems in current affairs in the spirit of *Scientific American*. Topics that have been covered include:

- Scoring streaks and records in sports;
- Health risks of electric and magnetic fields;
- The effectiveness of aspirin in preventing heart disease;
- Statistics, expert witnesses, and the courts;
- The undercount problem in the 1990 U.S. Census;
- Extraterrestrial communication;
- The use of DNA fingerprinting in the courts;
- The story of Deming and quality control;

- Randomized clinical trials in assessing risk;
- The role of statistics in the study of the AIDS epidemic; and
- Evaluating a needle exchange program to prevent the spread of AIDS.

As you can see, *Chance* magazine is a wonderful resource for teaching statistics and should certainly be in every college and university library.

Our original idea was to choose three or four of these topics and to try to understand the problems studied, the statistical or probabilistic issues involved, and the techniques that have been used to solve these problems. We soon found that many of these topics, as well as other equally interesting Chance course topics, were regularly reported in daily newspapers such as the *New York Times*, the *Washington Post*, and the *Los Angeles Times* and in weekly journals such as *Science*, *Nature*, and the *New England Journal of Medicine*. Thus we have used a blend of current news and articles like those in *Chance* magazine for our course.

The Chance course is not meant to be a substitute for a basic statistics course or for an introductory probability course. Its purpose is just to make students better able to understand and make critical judgments about reports in the media that involve probability or statistical concepts.

The first thing we discovered was that we ourselves needed some background information on topics as technical as DNA fingerprinting, and we needed some help finding current news articles. For this reason we established the CHANCE database on gopher on the Internet to make available to others our experiences in teaching current topics as well as to provide background material for these topics. We also started an electronic newsletter available by e-mail called Chance News. Chance News abstracts articles from the last couple of weeks' news that might be useful in teaching a course like Chance. Anyone interested in obtaining this Chance News can request it by electronic mail from dart.chance.edu. You can also obtain current and back issues of Chance News from our CHANCE gopher database.

I think that the ability to share information on the Internet will greatly enhance the teaching of statistics in the future. For example, let me show how easy it is to get our Chance News from gopher. The procedure is a little different for each machine, but I will illustrate it in terms of a Unix machine. Assuming that gopher has been installed by your system manager, and this is the case at most institutions, you just type

```
gopher chance.dartmouth.edu
```

You will then see a list of items, one of which is CHANCE Database. You move the pointer to this and then you will see:

```
                CHANCE Database

---> 1. Search CHANCE database <?>
     2. Welcome.
     3. Chance_Course/
```

4. Chance_News/
5. Chance_Profiles/
6. Current_Chance_News.
7. Fair_Use/
8. Other_Gopher_Servers/
9. Texts_of_Articles/
10. Tools/

If you want all the articles on a given topic such as SAT scores, you can use the search option. If you want the current Chance News, just move the pointer to Current_Chance_News and you can scan through the abstracts of recent articles. For our own class we browse through these abstracts to find one that might be interesting to discuss in class. For example, we might choose the recent *New York Times* article about the study showing that homosexuality might be partially a genetic trait. Here is the abstract of this article:

> Report suggests homosexuality is linked to genes.
> New York Times, 16 July, 1993, A,1.
> Natalie Angier
>
> An article in the 16 July issue of *Science* reports on a study led by Dean Hamer that suggests that sexual preference has a genetic component. The study began by looking at the family histories of 114 homosexual men. The researchers discovered a higher than expected number of gay men among the men's maternal uncles and male cousins who were sons of their mother's sisters. This suggested a gene or genes that pass through the maternal line and therefore through the X chromosome.
>
> This led them to study the genetic material from 40 pairs of brothers who were gay. They found that in 33 of the pairs the brothers displayed the same cluster of five markers bunched into a small region on the X chromosome. Since there should be a fifty percent chance that two brothers share the same allele by descent, this finding is regarded to be significant.
>
> The authors are careful to say that this study will have to be replicated and, if correct, is surely only one part of the answer, and so on. A similar finding two years ago that suggested an anatomical difference between the brains of gay and straight men received a lot of publicity and led to heated debates, mixed emotions, and so forth.

We would then provide the students with the full text of the article to read and ask the class to break up into groups and discuss for 20 minutes or so three or four questions concerning the study. For example, we might ask:

1. How convinced are you by this study that there is a genetic component to homosexuality?

2. Does this study tell you anything about the proportion of the general homosexual population or the non-homosexual population that have these "gay genes"?

3. What further studies will be necessary to clarify the extent of a genetic component of homosexuality?

4. What legal and other implications would establishing a genetic connection for homosexuality have?

When students discuss current issues in small groups they get totally immersed in the subject, and the most successful discussion questions develop a high level of noise and excitement. We are constantly surprised by the new and interesting ways that students look at a problem when they have not first been told how they should view it. It is an interesting challenge for the instructor to try to respond to the diverse ideas developed by the groups.

If we wanted an article that would lead to a more extensive discussion, we might choose from the Chance News the article on the recent supreme court decision on determining criteria for admission of scientific evidence in the courts. We could discuss this in terms of the current controversy over the use of DNA fingerprinting. Here the students can see arguments about statistical conclusions in a life-and-death situation. By using the search capabilities of our database we could search for previous articles on DNA fingerprinting, and we would find reports on a number of legal cases in which the following statistical problem played a key role.

A locus used in forensic DNA fingerprinting has a number of repeats of a particular sequence of DNA letters. For example, AGGAGGAGGAGG might be a particular allele for a locus having four repeats of AGG. For the loci used, there is a great deal of variation between individuals in the number of repeats. These numbers can vary from a small number to several thousand. Each individual has two such alleles, one obtained from the mother and one from the father. When there appears to be a match for these alleles between the DNA of the suspect and the DNA provided as evidence, how do you evaluate the chance of such a match for a randomly chosen person in the population? The Hardy-Weinberg law says that if our population has random mating and no immigration or migration, then the alleles in an individual can be regarded as the result of independently choosing two alleles according to the proportions of each allele in the whole population and we can use the infamous product law for independent events.

The FBI and commercial companies that do this kind of analysis invoke Hardy-Weinberg applied to several loci to get remarkably small probabilities for a match. How can they justify this independence? Obviously, if they were looking at alleles that determine the color of your skin, they could not. But the loci used for DNA fingerprinting are taken from the "junk" part of the DNA that is not thought to serve any particular purpose. If it is just part of your junk DNA, why should Hardy-Weinberg not apply? After all, your parents hardly decided to marry based on the junk part of your DNA that has no known purpose. But even then, might different ethnic groups have different distributions for the alleles? What statistical tests can be used to justify the Hardy-Weinberg assumption? What about the measurement errors in establishing the number of repeats? These are the kinds of questions that are being raised in the courts in trials using DNA fingerprinting. If you ask students to imagine that they are testifying in court, they will get a good chance to think about statistics in action.

There are many other potential uses of the Internet for more advanced students of statistics. For example, students might be encouraged to participate in one of the statistics discussion groups. Previous discussions for these groups are archived on gopher, and so you can again search for all discussions on a given topic. For example, if you search on the archives of STAT-L or EDSTAT-L for messages on Simpson's paradox, you will find some interesting real-life examples described and referenced. You can also get data sets and information about statistical software from the STATLIB archives, and you can even read in those archives the current *IMS Bulletin* significantly before it appears by mail. All this and more can be found by pointing your gopher to the electronic mail address jse.stat.ncsu.edu, where you will also find the new electronic *Journal of Statistics Education*.

Journal of Statistics Education

---> 1. About the Journal of Statistics Education Information Service.
2. Methods of Access.
3. Index to the Information Service.
4. The Journal of Statistics Education/
5. EdStat-L Archives/
6. Software Tools and Information/
7. Other Discussion Groups/
8. Other Services (Census, Statlib, etc)/

In item 4 you will find the first issue of this new journal, and you can browse through it to find more information about learning in groups in an article by Joan Garfield, and an interview with Fred Mosteller explaining the evolution of his work up to the present time that finds him deeply involved in interdisciplinary work in the finest sense. Robin Lock's Datasets and Stories section of the journal will be a valuable resource for new ways to introduce interesting data into our courses. If you find something you like, just print it out or even mail it to a colleague or student.

Previous speakers have explained the need to get students to think about the issues that cross disciplines. One place to start is to try to explain what you read in your daily newspaper. That has been our approach, and it has been a lot of fun.

Modernizing Graduate Programs in Statistics—Case Study

Prem K. Goel
Ohio State University

First, let me thank CATS for inviting me. About two years ago this topic arose at the summer CATS meeting, and there was a lot of enthusiasm about doing something in this direction. Recognizing that change happens very slowly, if we start working toward it now, maybe the curriculum changes will be in place in most programs by the year 2000.

I will try to avoid repetition of ideas, although many of my perspectives have much in common with those of previous speakers. I will concentrate on graduate curriculum, not because undergraduate programs are unimportant, nor that service courses are not important for the future of statistics, but because when I thought about this symposium in those initial CATS deliberations, I was mainly thinking about the future graduate curricula in statistics. Later, I will say something about what we have done in terms of service courses at Ohio State University (OSU). I will not give details, but rather will describe what we are doing in general.

Earlier this year while visiting Cleveland, I read an interesting story in *USA Today* about total quality management (TQM) awards given by Rochester Institute of Technology for the best TQM project in industry. The story described a paper mill company in southern Louisiana that won the award. Their motto is "In God We Trust; All Others Must Have Data." They developed this motto because they had previously had many problems in their plant, and realized that the only way they could learn about what was happening in the plant was by going out in the woods and collecting data. The perceptions of the managers were nowhere close to what was actually happening in the plant and the production process. That is where the motto came from. But as far as statisticians are concerned, the problem is, What if you do not know how to use the data?

The reason we ought to be doing ever more interdisciplinary education in our programs is, as many people here have already noted, that our science needs the nourishment provided by interactions with other sciences. The development of statistical science is driven in large part by problems arising in other substantive disciplines. Today's customers, industry and the government agencies, demand that our master's and PhD degree holders be able to work well in research teams. That means we have to design our curricula to meet such demands and instill such qualities.

The three main users of our product, academia, the government, and industry, are all demanding that our students become super-statisticians. All want individuals who have strong theoretical knowledge in statistics and probability; who have strong statistical computing and simulation skills; who know how to model substantive complex problems; and who are adept at using a wide spectrum of the methods in the ever-growing data analytic tool box in statistics; who have strong communication skills; who are nimble problem solvers; who are able to formulate real problems in interdisciplinary projects and not lose sight of those problems; who are well trained in the art and science of consulting; and many more things. We are talking about producing a student today who knows everything about everything, and that is not easy to do.

It is clear to me that what we have been doing in statistics education up to today has to change because the world has changed around us. It is no longer the environment it was 25 years ago when the last major curricular changes took place in statistics. I am not as sure as to how we should change. We at Ohio State are still experimenting and learning what works and what does not, as are many other institutions doing similar things.

The need for change is equally true for undergraduate education, and not only for graduate education. Changes are also needed in undergraduate service education, but it is much easier to reinvent or redesign service courses partly because there is little vested interest in the existing service courses. That is not true, however, at the graduate level, where some people have been teaching courses for the last 20 years. If you now tell them, "You have to change your course because it is not relevant anymore," many of your colleagues may balk because it is in their nature to be conservative. This imbues the system with an opposing inertia that simply cannot be changed overnight. Change does not come easily and does not happen by itself. As Ed Rothman said, we must take the attitude that if we can change even ourselves, perhaps that would be a contribution to the profession. Hopefully, everybody will over time re-examine and implement the changes that ought to be made in statistics teaching and research.

Most statistics departments are not as fortunate as is Carnegie Mellon's. At CMU there is a fairly homogeneous group of people, and they have found a common purpose. Many older departments have people with very established ideas. For a department chair or a curriculum committee chair to work with those people and gain their support to look at curriculum very carefully, redesign some courses, drop some courses, and create some new courses, is a fairly tough job. It is especially so when the reward system does not reinforce doing that.

Basically, the reward system is predicated on how many papers one published last year. But if that is all that is rewarded, faculty members do not want to put much time or effort into teaching, partially because there is not much benefit in it for them. As departments, we have to think about how we can change our reward structure so that teaching becomes an important activity on a day-to-day basis. Such changes will happen because there is demand by the customers for us to change.

The ideas presented by the other symposium speakers merit serious discussion. What is important is for us to go back to our own departments and start talking about these things with other colleagues, and find ways to assimilate some of these ideas into our own programs. Again it will be a very slow process, and will not be easy. However, if we can make one convert at a time, perhaps we can succeed.

For the last four to six years, the OSU statistics faculty have been discussing what is wrong with what we do in the classroom, how our students learn or do not learn, and whether or not they are able to put their knowledge together in solving problems. The latter is one area in which we notice particular difficulty. Sometimes after a student has taken the mathematical statistics sequence, the probability sequence, and all the applied courses, and is then given a problem in a consulting setting to solve, he or she do not know what to do with it. Some of our faculty believe that may be caused by the way we teach the courses. We teach all these methods courses in isolation and do not build one course on the other. That may be why the students are not learning how to synthesize.

Many trained statisticians act more like technicians rather than scientists, and it may be in part because various methods courses are taught in isolation from each other. Phillip Ross remarked that if you have a hammer, you view everything as a nail; we may have to change some of our teaching to address and counter this tendency. Edward Rothman described statistics as technology and not as science or an art. I certainly do not agree with that, even though Fisher may have said it, because those may have been different times. If you want to be a full partner in the future development of science, you have to be a scientist today, and not just a technologist, because scientists will not accept you as a full partner if you are merely going to be concerned with a little piece of the problem. To work as full partners with scientists, we have to act like scientists and learn like scientists, too.

I believe that our students do not learn how to solve substantive problems, but are essentially looking for a tool that is already in their tool box to fit to a problem. That may be why there is so much emphasis on linear models when many real-world problems are not necessarily close to linear models. Even when dealing with nonlinear models we seem to have missed the boat completely, because we have simply tried to look at the series expansion of the nonlinear expression, and tried to fit a linear model; that is, basically, we do not focus on what actually is taking place. This context of solving substantive problems is one in which things need to change in our teaching.

To address this problem, the OSU faculty agreed that something needed to be done in, as a start, at least the PhD curriculum. The master's program got second priority because its functioning was viewed as acceptable. It was decided that a new course would be developed. In so doing, we have made substantive changes over the last four years in the Ohio State statistics PhD program. First, many graduate students are needed to teach in their very first quarter at OSU because the new resources recently obtained from the university for teaching the general-curriculum statistics courses require a large number of teaching assistants. Consequently, we support about 75 teaching assistants in our program. These people are needed in the classes in the first quarter. With the help of the administration, we arranged for these people to receive training in the summer quarter. We asked them in the first year to come to the university in the summer quarter. We support them for this with funds provided by the university. In that summer quarter, they learn how to teach laboratory courses, how to use the Data Desk or Minitab or whatever other tools or concepts will be discussed in the service courses. They also take what is basically a refresher course in mathematics, because many of them forget calculus. As undergraduates, they take the calculus in their first or second year and thereafter do not use calculus (in contrast to engineering students). So by the time they arrive in their first year of the graduate program, they have forgotten their calculus and do not know such things as how to transform variables, even with only two variables.

When they come, we ask them what courses they have taken, and what concepts they remember, and then we try to fill the gaps in that summer quarter so that, on the one hand, they are ready to take on the teaching assignments and, on the other hand, they are also ready for the first mathematical statistics course that is given in the fall quarter.

The change in our program is not in terms of the requirements, but rather in how the methods courses are presented in the first year. The students are still required to take a three-quarter sequence in a probability-mathematical statistics course and a three-quarter sequence in the first year in real analysis. That is sometimes deadly, because by the time they have taken the third course in real analysis, they might decide not to continue the program because they thought that they had not come to learn real analysis. However, they have to go through the program, and sometimes they are taking this course with all the math majors, too. Sometimes they cannot compete as well, or, not knowing that they will need it in the second or third year, they lose sight of why they are going through the real analysis sequence; they may decide to convert to a master's program and go out to work. Sometimes it is easier to get a master's degree and go get a better job than to put in six years for a PhD program and not know what will happen afterward; with today's job market in academia, we lose some of the students to a master's program, and, at least in the short term, they think they are better off.

The biggest change in our program was the following: In the PhD program, students used to take all the methods courses in the first year, such things as regression, analysis of variance, design of experiments, time-series analysis—you can as well make the list. Second-level courses were for master's and PhD statistics students, and also for some of the PhD students from engineering and other departments.

We decided to introduce a first-year sequence called Introduction to Statistical Practice. That is a three-quarter sequence of courses in which students learn about these methods through real problems. Two years prior to the introduction of the course, before it was implemented, we asked the faculty which of them would like to help team teach if there were to be a course like that. Three or four people volunteered. We then gave them some release time from their teaching duties in those two years prior to the introduction of the course to develop materials, to look for data sources and find real data, to think about how to structure the course, and things of that sort. This was to avoid having all the students in year one being guinea pigs.

The approach is to teach substantive problem solving through various large and small data assays. The class format is one of less lecture and more open discussion. Groups of five students are formed who will work together during the whole quarter. More attention is paid to asking questions and identifying what the real questions are, than to just solving the problems.

All the things we have been hearing the last two days—formulating the problems, identifying what the questions are, raising questions about the data, finding out what the data are about, determining how the measurements came about—are done in this course. This permits the first discussions in this sequence to address the scientific method, team work on formulation of problems, the art of raising questions, and the use of computing tools. Once they have looked at a problem and the questions have been formulated, then the instructors say, "If you want to solve this problem, how do you answer the questions raised in this discussion using appropriate tools?"

That is how we introduced the methods in this statistical practice course. Methods are not done in great detail in the course because there are too many things to discuss: about 12 real problems had been picked through which they were trying to introduce methods. Consequently, these students do not necessarily learn all about regression modeling, or all about detecting outliers, because they cannot go into that kind of detail in a course that is to give them an overview of the discipline and an overview of how to solve problems.

The idea is that once they have taken this course in year one, and have also gone through the mathematical statistics sequence in year one and have taken the probability sequence in year two, they will start taking some of the advanced topics courses in year two and year three in which they can choose, if they want, to learn multivariate or linear models or design of experiments or nonparametrics. They can choose which topics they want to learn more about. Hopefully, in this first-year Introduction to Statistical Practice course, they will develop enough curiosity about learning some of the things on their own, or will learn some of these things when they are taking a consulting service course, which is also part of the program.

Every student must take two quarters of a consulting course. The first quarter presented in the classroom includes discussion of the art of consulting, as well as how to deal with the client, report writing, and communications. In the second quarter each student gets involved in a real consulting project that is substantive in nature; students work for almost two quarters with one client on a project.

When you run a consulting service, you cannot tell clients their problems are not important. Each person's problem is important to him or her and therefore should be to you, too. Some of the problems that come to the consulting service are of the kind that can be handled in a couple of meetings; in such cases, we hope that the next time around, if clients have had a good experience from getting this help from us, they will return with a much more substantive project. That is how students also learn to get involved in both short-term and long-term projects through the consulting service.

That is the small innovation we have tried to implement in our graduate statistics program at OSU. The statistical practice course first covers problem formulation, and then techniques to solve problems are introduced through that process. In this give-and-take when the students are tackling these problems every day, five people work with each other. They return to the same techniques several times, because some of the things are used in more than one problem. The focus is taking a synthetic approach to scientific problem solving, not merely introducing methods or a list of methods in the course. Sometimes we have to team teach these courses because one or two individuals may not be able to cover the spectrum of statistics in an advanced applied course. I like this idea of team teaching. I think that bringing people together to team teach these new courses is one way to induce some of the desired changes.

Currently, two people who are experienced in real-world data analysis are teaching the course for the first time. They learned a few things this first year. First, first-year graduate students often want everything to be very structured, perhaps partly because it is such a big transition from an undergraduate program. They want to know what the course is going to cover and how grades will be determined. Grading is very important to them, and if you tell them, "A grade is not important; what is important is the material you should learn from the course," that does not resonate very well. That is especially so if they think you are trying to be vague about the grading system. This was in part the reason that in the first quarter of the course, many of the students were not happy with it. The course material focused on four problems; at first they merely discussed the problems, and students would say, "I have not learned any method yet; what is going on?" It did not entirely satisfy them to be told, "You will learn methods later; we first have to discuss how to ask the right questions before we address how to solve those problems."

This made the first quarter very difficult. Students were complaining that they were not learning anything that they had expected in terms of, say, regression modeling. In reply, they were told that the first quarter introduces computing and discusses how to raise questions, and the second quarter presents how to solve those problems. We have made a commitment to the two faculty members that they will be teaching the course for the next two years and thus learning from their own experience. They will be talking to other faculty members about it. At the end of two years, they will hopefully have gained experience and understanding that can be further discussed as a case study.

We at OSU consider the consulting service not just as a source of revenue for the department but also as a source for problems. Students who are assigned to substantive projects sometimes also end up getting a thesis problem out of the experience. There were about seven or eight theses in the last four years that came out of projects out in the field, where faculty members were involved on a long-term basis. These are instances of students getting involved enough to extract a problem that was good enough for a PhD thesis, and thus contribute to methodology development. I think it is very important to have a consulting service in any statistics department's program.

There are often difficulties in running a statistical consulting service. Not every faculty member is supportive of the effort because, for instance, some may view consulting with clients as beneath their dignity. That happens in many of the bigger departments. What the department's purpose is may not be agreed on by all the faculty, because their view depends on what the faculty think is important and on what they perceive as the basis on which they will be reviewed for promotion and tenure. I believe that the six-year time line for an assistant professor's getting promotion and tenure is sometimes detrimental to the interdisciplinary nature of statistical work. It can take several years for a person to become productively involved in some of the big problems; if other department members do not see any publications from such a person for two years, it can raise concerns over what is happening with that person. I think our job is not only to convince the deans and the provost, but also to convince ourselves and our colleagues that patience is needed. If faculty do not publish something in year one or year two, but we know they are very deeply involved in something that is going to produce results three years down the line, we have to keep that in mind and take it into account, or else we will discourage that kind of involvement for our junior faculty. Junior faculty sometimes need to be protected against that, because if they invest their time doing interdisciplinary work and get "passed over" for not publishing enough papers, it will be a major problem.

We also try to place many students in summer internships in industry and government and have had students go to several companies in Cleveland, Cincinnati, and Columbus. Students have been placed in the Bureau of Census and Bureau of Labor Statistics, and in other government agencies. When they return, they are very pleased because they have learned many things they would not have learned at the university. Internships are very important.

I want to appeal to our industrial colleagues to offer more opportunities for graduate students to be interns, and perhaps also offer faculty exchange programs. If those opportunities are not provided, despite all the talk we will not achieve what we want to achieve in interdisciplinary education. Industry must also be a full partner in the process, because otherwise it will not be possible to implement these interdisciplinary facets.

Before I close, I must acknowledge that some problems exist in the basic statistics educational curriculum at Ohio State. One is that not all faculty members are convinced of the effectiveness of this modern interdisciplinary approach. The problem is that some people do not ever want to change; they simply have developed their habits and that is it. Consequently, any attempt to change that faces opposition. Such people do not believe in this interdisciplinary approach because they never learned it that way yet can solve problems now, and so they wonder why students now cannot do the same thing that they did when they were graduate students and then develop over time. However, as Jon Kettenring said yesterday, industry does not have the time or resources to provide on-the-job training, and in previous times there was not this kind of pressure on the outgoing graduates to know everything when they get out. Things have changed. With so many more people in the statistics profession, perhaps supply is overtaking demand, and we will therefore have to be more cognizant about what our students should know before they graduate.

Another problem is the vested interest individuals have in the courses they have been teaching for the last 20 years. It is sometimes a big problem because if they are suddenly told, "By the way, we will not offer this course any more because it is not important," their reply might be in effect, "You cannot do that to me; I am a senior faculty member in the department, and you had better take care of me or I will create trouble for you." That is a somewhat facetious possible reaction, but there will be problems of that kind arising when you want to change curriculum. One has to work with everybody, decide on what is important, perhaps put some priorities on certain aspects, and try to change the course so that it incorporates the new higher-priority things, rather than simply cutting the course. It is important to avoid hurting people's egos.

One problem with the Introduction to Statistics Practice course is that no book is available yet. We intended to use two or three books as a collection to be read by students, but having students buy three books for a course is expensive these days. So we have put a lot of data sets on our own departmental machine that students have access to, and for now we provide a lot of handouts for this course. The hope is that after three years, the professors who developed the course will have a course format that can be made available at least in manuscript form for other people to use.

There is also a problem in that some of the students do not seem to be comfortable with a very unstructured kind of course because they are accustomed to a more traditional, series-mode teaching style that presents a problem, the solution, the next problem, and the next solution, rather than an appeal to try to learn how to formulate problems first. Making that change is not an easy thing for some students to do, and they do complain about it.

Still, the biggest obstacle is that a substantial number of faculty members do not want to work on particular private projects. They feel that theory is more important than are applications, and question why they should "waste their time" working with data. That, I would venture to guess, is a problem in many big, established statistics departments. Reiterating, to solve that problem, we will have to think about the reward structures. At Ohio State faculty are informed that those who help in the consulting projects have something in it for them. Those projects are usually funded ones, and when the money comes to the consulting service it basically is money that can be used for whatever educational purposes we want. It is not taken back by the administration, and so it is a department resource that can be used to buy some books or provide an extra trip to go to a meeting for those who help out in the consulting. We hope that through these kinds of little inducements we can attract more faculty. Some of them are very willing to help on any problem; others simply do not care about consulting. We believe that, over time, through the recruitment of more and younger faculty, this problem of not wanting to participate will diminish on its own.

Respondent

What Is Interdisciplinary Research?

Stephen E. Fienberg
Carnegie Mellon University

Only a week or so ago, I was very concerned about having to respond to the case studies because I had received no papers to read, despite promises that papers would be available in June. I became even more concerned as the date of this symposium got closer, and I arrived in San Francisco with nothing of my own on paper. I finally acquired a copy of the third talk and made some notes, trying to anticipate how the presentations would turn out. In the end, I have been pleasantly surprised, because the case studies and the discussion during the past two days have in fact fit into the framework that I laid out in my notes prior to the start of the symposium. Thus, my response is a partially planned presentation and discussion, but linked to topics that have emerged over the last two days, and in particular to issues raised in the three case studies.

To my mind, there were actually four case studies; the first was presented yesterday, with John Lehoczky describing the approach that my colleagues at Carnegie Mellon have taken, and I will make reference to it as well, partly because I am familiar with it and linked into it.

What Is Interdisciplinarity?

As I read preliminary versions of yesterday's presentations, I decided that there was some confusion between three different terms: "interdisciplinary," "multidisciplinary," and "cross-disciplinary." Thus I begin my response by focusing on what I think interdisciplinarity means, and then I will link my notion to the presentations we have heard today.

When I speak about multidisciplinarity, I mean people coming together, representing different disciplines, and somehow trying to work with one another but not necessarily changing their approach or adapting to the knowledge base or techniques of the other disciplines. This is not what I plan to talk about today.

When I speak of cross-disciplinarity, I have in mind people from one discipline, say chemistry, choosing to work in another discipline, say biology, and bringing to bear the tools of their original discipline in the new one. I think that this notion is closer to what we have been discussing, but it still falls short.

What is interdisciplinarity? In part, interdisciplinarity to me means people from two or more disciplines coming together, learning about the ideas and research in each other's fields, and then developing new ideas and research that meld those from the original disciplines. For the statistician attempting to engage in interdisciplinary activities, the first step is to learn what researchers in another discipline do and how they do it.

I have listed some of the things researchers in other disciplines do. They think, they speak, they have their own language, they theorize. And theory itself means something different in each discipline with which I have ever had some involvement. They measure things. There is a form of empiricism and a style of measurement in each discipline, and tools for measurement. And then they evaluate empirical evidence, often with their own analytical and quantitative tools as well as with the standard statistical methods with which we are quite familiar. Statisticians must take these activities seriously if they plan to work with people from another field.

Four goals that for statisticians who are going to engage in this kind of interdisciplinarity include, but are not restricted to, the following. A primary goal must be that you plan to contribute to the other field—that is, that you as a statistician have something to bring to your collaboration.

At the same time, if you are engaging in scientific collaboration you hope not only to contribute to the other field but also to draw something back to the field of statistics, and I distinguish between two kinds of impact on our field. First, one may draw back new statistical methods, or at least variations on methods that have been appropriate for the questions that you are investigating in the language of the other field involved in the collaboration.

But then the statistician occasionally takes the methods as embedded in that field, steps back, and thinks about them, linking them to methods from other disparate fields. The resulting integration of ideas and methods becomes applicable more generally. A unique feature of statistics as a discipline comes from the ability of statisticians to generalize from multiple disciplinary settings and to devise methodology that has broader applicability than to one field alone.

Something that we have not heard a lot about today and yesterday is my fourth category of goals for interdisciplinary collaborations. If you are really doing interdisciplinary work involving, say, geology and statistics, and if you are really successful and the problems are important, then what you may end up with is a melding of techniques, ideas, and language that do not fit naturally in either geology or in statistics, but indeed form what those who do interdisciplinary research refer to as an *interdiscipline*.

The exciting projects that I have been involved with are ones that take on that interdisciplinary character and go far beyond even a sustained consulting opportunity.

I thought that some personal examples from the various speakers at this symposium would help to describe interdisciplinary activities to others and explain why we as statisticians find them exciting. In that sense, I have been disappointed, especially because I know that many of the people in this room are working on substantive problems and making important contributions. The trouble is that we tend to speak in a meta-language. We step back from what we as statisticians are doing when we engage in interdisciplinary work, and thus we fail to convey to one another the excitement that goes with it.

That excitement of discovery and creativity is what I think all of us really want to see in the statistics curriculum.

Examples of Interdisciplinary Activities

I have listed three topics that are examples of interdisciplinary research that I have been engaged in off and on over the last little while:

- Cognitive aspects of survey methodology,
- Statistics and the law (including the use of DNA fingerprinting as evidence), and
- Designing and evaluating bilingual educational programs.

I will talk about the first at length because I suspect it is the one you may be least familiar with. The second is something that has been mentioned several times in the last two days in different guises and for which I will provide only some references to my own work (see Fienberg, 1989, 1990; and Fienberg and Kaye, 1991).

The third topic was a project of a panel of the Committee on National Statistics (see Meyer and Fienberg, 1992) and focused on a critical public policy issue of how to educate students with limited proficiency in English. The panel involved statisticians, educators, linguists, and survey specialists, and it evaluated two major studies of different approaches to bilingual education. Now let me return to the first topic: cognitive aspects of survey methodology.

In 1980 I was engaged in work with colleagues studying crime victimization. This was interdisciplinary research, and it involved criminologists, social psychologists, sociologists, and statisticians. We were exploring how to understand victimization patterns in the United States, using data that had been collected as part of the National Crime Survey—a household survey launched in the 1970s under the sponsorship of what was then the Law Enforcement Assistance Administration and implemented by the Bureau of the Census. With the creation of the Bureau of Justice Statistics in 1979, there was an effort to reexamine the survey and, in particular, to try to come to grips with reporting bias. Virtually every expert on criminal victimization seemed to believe that we were not measuring enough crime, that is, that there were more crimes being committed than the respondents to the survey were reporting. This was bias that, everybody perceived, was associated less with sampling procedures and many of the traditional sources of nonsampling error, than it was with the questionnaire—the fundamental design of the instrument used to measure victimization.

At a very early stage in our work, somebody said, "Maybe we do not know how people think about the questions that are being asked when interviewers go out in the field." And then someone else said, "Who knows about how people think?" "Well, psychologists worry about that," the first person responded. "Wouldn't it be interesting if we did something multidisciplinary and took a group of psychologists, a group of criminologists, and a group of people actually interested in the National Crime Survey and put them in a room and let them talk!"

In fact, we did just that in the form of a two-day workshop. Mainly what resulted was babble, because the psychologists had their language and they wanted to tell us about what they did, but only in their own language; and the criminologists had their language and the survey people had their language. What was interesting was that the three groups often used the same words, but they did not mean the same thing.

Nonetheless, there was an interesting dialogue. What happened after a while was that a couple of people went off and they actually talked about a problem that lay close to the interface of both their interests. Then a few more did the same. The result was that some of us got quite excited about the possibility that there was something more in this enterprise, even though this particular workshop was not coming off all that well in that we were not learning how to redesign the National Crime Survey.

A few years later, the Committee on National Statistics (the other statistics committee at the National Academy of Sciences) organized a week-long workshop. We selected people from the fields of cognitive psychology, statistics, and survey research and asked them to explore in very different ways what each of the fields did, what they had in common, and what they could bring to one another in the form of research ideas and research opportunities. Further, we commissioned special background papers to try to draw together research ideas at the interface that would be explored at greater length.

Finally, we went to the people at the National Center for Health Statistics who did the National Health Interview Survey, and we asked them if we could use their survey to focus our attention.

Because we wanted all of the participants to have some common knowledge, we arranged for each one to be visited by one of the interviewers in the National Health Interview Survey (NHIS). That is, we arranged for somebody who would normally go out and interview a family somewhere in the United States to come to each of our homes. My wife, my older son, and I sat in the living room of our house and went through the hour-long interview for the NHIS. The interviewer did not know why she was coming to my house. She did know, however, that we were not part of the regular NHIS sample. In planning for the workshop, we also sent the interviewers out to carefully selected households—not because we knew what those in the households had to say, but rather because we needed their cooperation—and we filmed the interviews.

Later, during the course of the workshop, we actually got together in the evenings and watched the filmed interviews. We would be watching the videotapes and suddenly one of the psychologists would say, "My God! Look at what's happening here!" And then the psychologists would try to explain to us in psychological terms something that was happening in the interview. The governmental health statisticians thought of what was happening in the interview in very different ways, as did the statisticians from universities.

What we discovered was that the survey statisticians at the National Center for Health Statistics had been doing what we at this workshop were talking about earlier. They had a wonderful statistical tool—the household survey. It was like a hammer, and when you have a hammer, you look for nails. Well, they had a well-developed approach to survey-taking. They had a way to design questionnaires for household interviews, and they applied it to a series of issues on health. But when we watched the videotapes, what we learned was that the respondents to the health survey had stories to tell us about their personal lives and the consequences for their interactions with the health care system.

I have a vivid memory of watching a woman try to tell the interviewer that she had been in a car accident, and as a result of that car accident she had gone to see her doctor several days later. She did not leave work early any day, but she had had a headache and she had had "minor" symptoms. It was a full and consistent story, and it turned out that every other question in the NHIS questionnaire tapped into that story. The answer to each question that she wanted to give was related to this incident, and she had no natural way to explain this to the interviewer in the context of the NHIS questionnaire. So she would try to give the answer in her own terms, but this did not satisfy the interviewer, who would repeat the question, but slightly more loudly than before, and so on and so forth.

Over the course of that workshop, we began to have dialogues across disciplinary lines, we learned the language of others, we worked together, we developed teams looking at different aspects of that particular survey, and we developed interdisciplinary research projects. Jon Kettenring was the first to use the word "teamwork" yesterday, and teams played a very important role in that enterprise.

The participants left that workshop and started small research projects, and they did this in teams. The people in the statistical agencies actually brought in cognitive psychologists to work on survey problems. They began to use some of the tools developed by cognitive researchers—for example, something called protocol analysis.

There is a very long story associated with this interdisciplinary set of activities that I will skip over, but I can tell you that there are now cognitive laboratories in three of the largest statistical agencies in Washington. If you read the program of the American Statistical Association meetings that are about to commence on Monday, you will find several sessions devoted to this topic. For example, if you attend the session on the redesign of the year 2000 census, the user-friendly census form being described there is linked to work of the cognitive laboratory in the Bureau of the Census.

Cognitive aspects of survey methodology (CASM) is an enterprise that has brought together people from multiple disciplines, and they have formed what I describe as an interdiscipline. They now have their own language, their own dialogue — and they will probably have several journals before we are all done.

Statistics has had a critical role in this story. As the CASM interdiscipline evolved, we began to see an interplay between cognitive laboratory work, tightly and loosely designed experiments in the laboratory, and survey field tests so that we understand what happens when you make the transition from one environment to another — something that as statisticians we have learned about in other scientific disciplines. Finally, there have been formal tests, with well-designed embedded randomized experiments within surveys so that researchers can measure bias, measure change, and integrate ideas in lots of ways. (For further details see Fienberg and Tanur, 1989, 1994; Jabine et al., 1984; and Tanur and Fienberg, 1992.)

I did not actually plan to tell this story before coming here but I think it illustrates a number of themes evoked in this symposium, and it is a description of a very different kind of interdisciplinary activity that one does not normally associate with statistical education.

I want to go back for just a moment to my third example, bilingual education programs, mainly because of the word design. Ed Rothman said in the earlier discussion that most data are not designed; they come to us in different forms in the real world but with little planning. I do not think that came out in quite the way Ed intended. I agree with him that all statistical studies do not come to us in the form of a carefully designed randomized control trial, although perhaps it would be better if more did; however, almost every study is designed, by hook or by crook, and the problem is that most studies are ill-designed and have no possibility of yielding the answer that people set out to provide. In the Committee on National Statistics (CNSTAT) panel study to evaluate programs in bilingual education, we had to learn about what bilingual education meant, what bilingual education *theory* was, what happens in classrooms, and so on. Then we asked, "What does it mean to design studies involving different forms of bilingual education that will affect the development and achievement of students in the modern public education curriculum?" I will not tell you our answer, but I do commend our report to you.

How Do We Do Interdisciplinary Research?

My story on cognitive aspects of survey methodology was intended to give you the flavor of what is different about interdisciplinary research. I think it is very important to distinguish between interdisciplinary research and what has been referred to by Prem Goel, by Ed Rothman, and by a number of others as statistical consultation or collaboration. Even collaboration is not quite the same as interdisciplinary research, at least of the sort that I have described, but it comes close.

I want to reemphasize the design issue here, largely because when we talk about what goes into modern statistical education, we often are willing to take data from anywhere and think that that's a sensible exercise. I take issue with that. If you have an important question you have to answer, you must ask how to get data that will assist you. Only if you have thought seriously about data requirements and for some reason are not able to acquire the best data for the purposes, should you be prepared to ask: "Well, if we have not got these data, what can we say?" But I think that if you fail to ask the first question, the one of primary interest, then you have not fulfilled your job as a statistician. Therefore design plays a critical role when you are really collaborating with others, when you are really doing interdisciplinary work, as opposed to doing statistical consulting, when it is often too late to have any influence on what data get collected. While statisticians do not get to design all of the studies they collaborate on, they must learn to think in a design mode, rather than just in an analysis mode.

Building Applications Into Statistics Courses

Today's session was really about the case studies, about statisticians having attempted to build applications into the statistics curriculum. I want to pick up a few highlights and make some observations.

We actually had two presentations that focused more on the first course in statistics; and indeed, throughout the discussions at this symposium there has been a tension between how to teach the introductory courses versus where in the advanced statistics curriculum there is a place for serious applications. Ed Rothman described features of the first course developed at the University of Michigan, and Laurie Snell has told us about Chance, his new first-course effort developed at Dartmouth College, but transported elsewhere.

I think they are terrific examples. The detailed ideas and the illustrations Ed and Laurie have told us about are very good, and I think that the variation in approaches that Ed talked about are also important. I am not yet convinced about buying into the full framework and structure that Ed laid down, but I am excited about some of the components that he described and I suspect I will try some in my next undergraduate course.

I do think that, at some point, you need a philosophy and framework for statistics courses in a broader undergraduate curriculum or you get an unstructured course. And students tend to react poorly to unstructured courses. Prem Goel talked about the problems that his colleagues were having in the first go-round of a graduate course built around applications, and one feature the students reacted to negatively was the lack of perceived structure in the problems they were seeing. It is one thing not to structure the applications problems in the sense of predigesting them for the students, but you had better have a structure in what you are trying to get across to the students. Students need to understand that there are different kinds of statistical structures for different substantive problems, and I think that that is probably part of the problem the faculty members at Ohio State found in starting up their enterprise.

I cannot help but take a moment and talk about *Chance*, because Bill Eddy and I were there at the beginning. *Chance* is a magazine, not a journal, and when we created it, we intentionally designed it to have something like the look and feel of *Life*, *Time*, or *Sports Illustrated*. I am envious of the resources Laurie has been able to draw upon. When I last taught an introductory statistics class, I did not have the same rich array of materials at my disposal. The world of computing has clearly changed. His suggestion of getting the students engaged in looking at examples through the use of the Internet has given me some great ideas about what to do when I am back in an introductory class.

The last time I taught the big introductory statistics class at Carnegie Mellon, I did do some of the things that Laurie described. I began each lecture with a newspaper article or something in the news. Actually, I did not create this idea. Fred Mosteller used to do something like that in a different form back when I was a teaching assistant at Harvard in the 1960s. In my class, I actually encouraged the students to go out and find statistical material and bring it into class for discussion. But you have got to be very careful if you try this approach, because if the students do not have resources at their fingertips, what you can get is a lot of mush, and then it is very hard to respond positively to the students' efforts.

What we have now is an abundance of resources, and if you organize it just a bit and encourage students to reach into those resources, I think that there are exciting opportunities. I also wish Laurie had been around selling copies of *Chance* to the libraries when we were getting going back in 1987 and 1988, because that was and remains a difficult activity.

You can also link these things Laurie spoke about to videotapes. A couple of years ago, I taught a short course with a former science reporter for the *Washington Post*, Vic Cohen. Vic had written a book called *Numbers & News* (Cohen, 1989), and he was viewed by the science journalists as a statistician. What we did was to share with them several recent issues of *Chance*, and we were very fortunate because at that time David Moore had just put together *Against All Odds*, the wonderful Public Broadcasting System series.

Those kinds of resources exist, not just in *Against All Odds*, but also in many other places. We have classrooms that are fully automated and where it is easy to integrate standard lectures with computer-based and video materials. This requires preparation, but it is very exciting for the students and for the teacher. Parts of the *Against All Odds* lectures were drawn from *Chance* stories, and so Vic and I showed several excerpts as part of our presentation. Students and even journalists enjoy multimedia presentations.

We have also had some discussion at this symposium about focused method courses. The problem associated with those courses is mainly that the examples almost always are tied to methods introduced in the course. So the students know that when the smokestack data out of Daniel and Wood (1971) is used in a regression course, it is there for them to address with a linear regression model—even if they need to put in a nonlinear term later on, and even if all the assumptions are not fully satisfied. When they take these courses, students know the statistical tool to apply because it is the one that you taught last week just before you set the assignment.

Now, the difficulty comes when you want to study statistical problems rather than statistical methods. When you give students a problem, they do not have the hammer in their hand. They are not even sure if they should look for a hammer and nails. They do not know whether regression is appropriate, or categorical data analysis, or time-series methods. This is when they struggle. We have actually heard two different features of that struggle, and the Ohio State experience is telling.

That struggle involves students who do not necessarily have a tool kit to carry around to begin with and who thus have a real handicap. At least if they have regression analysis, then they can try it and find out that it does not work. But if they do not have at least one or two tools available to them, then it is very hard for them to address new statistical problems because they have not gotten into the mode of statistical thinking.

Yesterday, we heard from John Lehoczky, who described the Carnegie Mellon approach, in which the students come with at least a partial tool kit and are taught about how to draw in other tools. At the same time, however, they are reoriented to look at statistical problems, and they ultimately have to face the real test: taking a problem from beginning to end, working with others, and presenting the results of their effort.

I would emphasize the word "transition" here because I think that we do not yet understand—I speak for my colleagues at CMU as well as others that I have talked with—how you shift from one mode of statistical education to the other. You do not do it simply by stopping teaching statistics one way and starting to do it the other; you have to guide students into this second way of approaching problems involving, or crying out for, statistical thinking and methodology.

What Is MIUSE?

Now I want to ask a much harder question: What is this acronym—MIUSE—that expresses the topic of this symposium? Well, I can tell you what it is not; I am not sure I can tell you what it is. It is not the "alphabet soup" that John Lehoczky referred to yesterday, the collection of the latest "hot" computer-based data analysis techniques that fill our journals. These may be wonderful data analytic ideas—innovative, creative, and coming out faster than anybody can possibly keep up with—but they do not constitute MIUSE. It is not even dynamic graphics, whether they involve formal analysis or the kind of summary display that Ed Rothman was drawing our attention to a little while ago. It is not even statistical process control (SPC) and total quality management (TQM), a little more alphabet soup, although these are important tools for possible use in the enterprise.

How to Do It

The issue we have been getting close to, but not yet getting our hands on, is how to bring into the curriculum real interdisciplinary work of the sort described through my example earlier. Some relevant themes have come up over the last two days, and I thought I would at least make reference to a few of them.

The statistics faculty who teach about applied problems must actually do interdisciplinary research, or this just is not going to become a part of the curriculum that anybody will take seriously. One of the modes for accomplishing this is cross-appointments. This topic arose in a few presentations and then was either dismissed or at least set aside.

Over the course of my career I have had cross-appointments in theoretical biology, in social science, and in law. (My fellow responder on this panel, Ron Thisted, also has a substance-based cross-appointment at the University of Chicago Medical School.) A cross-appointment is not simply two sets of faculty meetings, two sets of students to be responsible for, and two department chairs or heads who determine your salary. A cross-appointment typically opens up for the statistician a wonderful source for problems for collaborative study. The person with the cross-appointment also needs to draw research problems back into the department of statistics and share them with others, not simply work on them in isolation. Statistics as a field has had a tradition of cross-appointments, but I suspect that we have had fewer than we should have had in recent years. There is a chicken-and-egg problem here. How do you get a cross-appointment in the law faculty if you do not have any expertise in the law, or at the interface of statistics and the law, that the law faculty members judge to be of value to them? So you have to work on this.

We have also talked about how you make room in the curriculum for interdisciplinary activities, and I mentioned in the discussion yesterday afternoon my view that we must look at the graduate statistics curriculum from a much broader perspective. Prem Goel raised the related problem of required courses in the graduate curriculum. If we ask about requirements one by one, then it is inevitable that some faculty members will support keeping each and every thing in the curriculum, without change, unless, of course, they wish to add additional required courses. All I can say is that we have adopted a much more flexible approach at Carnegie Mellon. This does not mean that we do not have similar kinds of pressures; they just manifest themselves in a different form.

There is another feature about how to do modern interdisciplinary university statistics education that we have addressed only in passing in this symposium, but I take the Carnegie Mellon example to illustrate my point. One faculty member does not take sole responsibility for the interdisciplinary part; a team of two makes it a little easier, but even two faculty members will end up having trouble if the class has more than a few students.

Essentially, what we are trying to do when we attempt to implement MIUSE is to teach our students how to work and think in an interdisciplinary environment. You do not do that by taking 30 students (over even 5 to 10 students) to watch a statistician work collaboratively with two or three colleagues. You must approach this either in very small groups or even one-on-one.

The CMU model of data analysis projects for our PhD students literally ties student and faculty together, working with someone else in a different field. There is this mentor-and-apprentice relationship that is an essential part of getting students going on applied problems and helping them through the inevitable rough patches. This is a very expensive way of educating graduate students, and we have to recognize that.

Laurie Snell mentioned something for introductory statistics courses that I think is very important for graduate courses as well: our students are not empty vessels. They often have an interest and experience in substantive fields other than statistics. I had a first-year graduate student who is my new advisee come into my office last week and ask, "Is it okay if I take an economics course? Do you think the department would be upset?" I replied, "Well, the one thing I can tell you is that the department will certainly not be upset. We may have to help you juggle various requirements, but all you have to do is explain why you're interested in studying economics and I'm sure that there will be support for your request. We think it's important that you have substantive knowledge, because the more things you know beyond statistics, the better you will be as a statistician." I am sure that, in two or three years, this student will bring his substantive knowledge about economics to bear on a statistics problem in a way that we did not anticipate. I believe that we must learn how to utilize such interests on the part of our students.

It is important that we remind ourselves as statisticians and our students about the rich tradition of our field involving people who work both on statistical methods and in other areas of science. I made only a brief list, in advance of hearing Jean Thiebaux make a related point yesterday: Pierre Simon de Laplace, Karl F. Gauss, Francis Y. Edgeworth, R. A. Fisher, Jerzy Neyman, William G. Cochran, Morris H. Hansen, Frederick Mosteller, and John Tukey. These are all people who drew strength for their statistical ideas from other fields.

In fact, when we talked about training problem solvers yesterday, I wanted to stand up and say that this is not a new idea. I would encourage all of you to read a wonderful paper published in 1949 in *Science*, called "The Education of a Scientific Generalist." In this paper (Bode et al., 1949), Hendrik Bode, Fred Mosteller, John Tukey, and Charlie Winsor described a curriculum that we today might label as the education of a statistical generalist, the student who is a problem solver and can move into new areas, learn the substance, work on difficult problems, and then move on, taking the lessons learned and putting them to work on fresh problems.

Finally, I note that we must be careful in articulating the goal for our students by bringing interdisciplinary approaches into the curriculum. It may be that even though you have faculty colleagues with a very rich tradition of working with others in multiple disciplines, not all of your students are going to feel comfortable in moving from area to area.

An outcome of modern interdisciplinary university statistics education may well be that we train a number of students who develop a primary interest in one of the interdisciplines, in one of the areas of application, but they may not move on to another interdisciplinary area at least for a long part of their career. This is different from the notion of training lots of people who will be busy solving problems in several substantive areas simultaneously, or possibly sequentially. I think we are much more likely to produce the first kind of student than the second kind.

Finally, I note that you cannot just expect materials for interdisciplinary statistics education to arise from your own resources. Even if you are a Fred Mosteller or a John Tukey, you are likely to not have enough resources at your fingertips to get every student going in a different kind of substantive problem. So we will need to rely on our colleagues for assistance, both local colleagues and others elsewhere. I think that we need a set of interdisciplinary research resources at an advanced level. One approach is to develop something like what is being proposed by the Institute of Mathematical Statistics, an *Annals of Applied Statistics*. This would be the kind of publication that would publish the case studies, that would get people excited about interdisciplinary work, that would provide the substance and the development of statistical ideas. These are things that statisticians are not accustomed to writing about in our typical journal articles. I hope that, if you have friends on the IMS Council, you will tell them that this is important for the field and that they should vote to support its development.

People have asked me why, if a topic such as cognitive aspects of survey methodology is so terrific, we have not introduced it in our graduate curriculum at Carnegie Mellon. I must confess that even we at CMU have to work on developing modern interdisciplinary statistics education. There is no course on cognitive aspects of survey methodology, no course in statistics and the law, and not even one in statistics and public policy. All of these topics are near and dear to my heart. But we do try to introduce some of our students to these ideas, even if it is in an informal manner. I note with anticipation that, in three weeks' time, I will begin teaching the data analysis course for our PhD students. I do plan to address there how to bring the kinds of interdisciplinary projects we have been discussing at this workshop into the mainstream of our graduate statistics curriculum. To me this is what MIUSE is all about.

References

Bode, H., F. Mosteller, J. Tukey, and C. Winsor. 1949. The education of a scientific generalist. *Science* **109**:553-559.

Cohen, V. 1989. *News & Numbers*. Ames: Iowa State University Press.

Daniel, C., and F. S. Wood. 1971. *Fitting Equations to Data*. New York: Wiley.

Fienberg, S. E., ed. 1989. *The Evolving Role of Statistics as Evidence in the Courts*. New York: Springer-Verlag.

Fienberg, S. E. 1990. Legal likelihoods and a priori assessments: What goes where? Pp. 141-162 in *Bayesian and Likelihood Methods in Statistics and Econometrics: Essays in Honor of George Barnard*. S. Geisser, J. S. Hodges, S. J. Press, and A. Zellner, eds. Amsterdam: North-Holland.

Fienberg, S. E., and D. H. Kaye. 1991. Legal and statistical aspects of some mysterious clusters. *J. R. Stat. Soc., A* **154**:61-74.

Fienberg, S. E., and J. M. Tanur. 1989. Combining cognitive and statistical approaches to survey design. *Science* **243**:1017-1022.

Fienberg, S. E. and J. M. Tanur. 1994. Surveying an interdisciplinary CASM. In *The Interdisciplinary Experience: Some Eclectic Reflections*. L. Salter and A. Hearns, eds. Montreal: McGill-Queens Press. In press.

Jabine, T., M. L. Straf, J. M. Tanur, and R. Tourangeau. 1984. *Cognitive Aspects of Survey Methodology: Building a Bridge Between Disciplines*. Washington, D.C.: National Academy Press.

Meyer, M. M., and S. E. Fienberg. 1992. *Assessing Evaluation Studies: The Case of Bilingual Education Strategies*. Washington, D.C.: National Academy Press.

Tanur, J. M., and S. E. Fienberg. 1992. Cognitive aspects of surveys: Yesterday, today, and tomorrow. *J. Off. Stat*. **8**:5-17.

Respondent

Ronald A. Thisted
University of Chicago

For the last day and a half we have been talking about change. Change involves risk, and risk involves the possibility of failure. For those reasons, it is uncomfortable to contemplate change. The possibility of failure makes us more comfortable with what we already do, and it makes us worry: if we do effect change, will we change in the right direction and with the intended effect?

My job as last speaker is not to be a discussant but to be a respondent, and so what you will get for the next few minutes are my responses -- my knee-jerk responses, if you like — to what has gone on for the last day and a half.

Jean Thiebaux's argument that statistics students should be grounded in philosophy is exactly right. Although she did not follow this idea very far, I think that philosophy has much to offer statistics. In particular, the area of metaphysics is important in philosophy. Metaphysics is the critical study of assumptions, of first principles, of where we start from and how that starting point affects where we can go.

This symposium has been exploring a particular model for change, a model for reform of statistical education, and I hope it is not too much to suggest what George Box constantly reminds us of: "All models are wrong" (Box, 1976) — not necessarily very wrong, but wrong just the same. And so I will devote some of this response to checking the model of education that we have been discussing -- to explore, if you will, some of the possible sources of bias in the conclusions we are drawing that depend in this metaphysical way on the way we framed the question.

In statistics, what we both learn and teach is that assumptions can be violated. A metaphysician would say that our assumptions can limit our thinking, and that being critical about our assumptions can sharpen our thinking. What follows are some of my reactions about the assumptions I see as being thus far implicit in our deliberations.

This symposium is based on the assumption that there is something wrong or, perhaps more charitably, that something is missing in statistical training programs, namely, this interdisciplinary component. What makes John Bailar's comments so disquieting is the suggestion that this assumption, which at the base is a very comfortable assumption for statisticians, is wrong, and that interdisciplinarity is not missing in our training programs but rather is missing in ourselves. Why is "interdisciplinary work" seen as bringing salvation; why is it viewed as a good thing?

I do not believe the real reason is because there are clients whose needs would be better served if statisticians did more interdisciplinary things. Although it is certainly true that statisticians have deficiencies related to failings in interdisciplinary work, fundamentally statisticians have an image problem. There is also a market share problem: the economists have already passed us, while at the same time we genuinely feel that statisticians have something to contribute that is real and important.

The heroes of statistics have been very broad in their approaches to the discipline. Steve Fienberg listed some of the heros; my list includes David Cox, F. N. David, R. A. Fisher, Frederick Mosteller, Jerzy Neyman, and John Tukey. All of these individuals are or were in their time taken seriously by non-statisticians in a way that I am afraid most of us today are not. But these are the data-savvy statisticians; these are the holistic statisticians, and, in truth, none of them are products of a modern statistics PhD program. So, perhaps, if the modern statistics PhD program is changed, more of these heroes of statistics or, as Brad Efron would prefer to call them, the wizards of science, can be produced.

They represent to varying degrees not so much what we as statisticians want to do but what it is we want to be. And this dissonance between what we want to be and what we now do is a proper note on which to return to consideration of some other assumptions. I next list some assumptions that have underlain the symposium discussions, not to suggest that these assumptions are wrong, but that they might be different.

One assumption is that the purpose of statistics training is to produce the right product for the right client. This business metaphor is useful. One can talk about the declining demand for the product, the increasing production costs and—what I believe is the greatest fear of directors of statistics programs—the possibility of excess inventory at the end of the year. But there are other potentially useful metaphors that should be considered. There is the Darwinian metaphor, for example, in which the purpose is not to produce a product but to replicate the species, and wherein the idea of incorporating interdisciplinarity is a way of evolving so as to avoid extinction. I think each of these metaphors puts a different emphasis on what our purpose ought to be. I like the business metaphor better than the evolutionary one, as Ed Rothman very clearly did in his presentation. The very first thing he listed for his curriculum was purpose; everything follows from some set purpose. This is also related to John Bailar's comments that perhaps our purpose and goals are wrong, and a clear vision is needed of what is the purpose of statistical training.

I am involved in creating a new department that I hope represents an interdiscipline in our medical school. The goal of this department is improving the health of the U.S. population. Maybe that is a little ambitious, but we statisticians have perhaps as good a shot at doing it as does Hillary Clinton, albeit in a different way. This is very much related to Jean Thiebaux's question, Who is the real client? Is it the future of science? Is it the future of our nation or our future global economy? I think all of those are important purposes and should be incorporated into our visions of who we as statisticians are and what we do; as Steve Fienberg has just suggested, at the very least we should maintain that statisticians should be contributing to other fields and by those interactions should be bringing something back to and informing the broad field of statistics.

A second assumption is that whatever is done in the statistics training program needs to be done in the context of a four-year PhD program. Twenty-five years ago the modal PhD program was a three-year program, at least in the rhetoric. Now the modal program is advertised to take four years, although the median time to degree in most programs is probably at least a year beyond what the literature says. "Four years" is an assumption. Perhaps making five-year programs a matter of course should be considered.

We can also look at other training models. The four-year program or the three-year program is not what is done in law, medicine, or business—or in physics or biochemistry, for that matter. In physics and chemistry, new recipients of the PhD go through postdoctoral training where they actually "learn their stuff." Once one gets a law degree, six months are spent cramming for the bar during which local law is learned, as is the material that was not taught in law school; one then joins a law firm as an associate for several years, and only after all of this is the person really considered a lawyer.

A graduate from four years of medical school does not go out and open a practice; instead, at least a three-year residency is taken. A graduate from medical school is not considered fit to practice medicine. The assumption that a new PhD should be fully trained, when nowhere else is that the case, seems a perhaps unrealistic one, and one for which the clients should take some responsibility. The clients have gotten off the hook a bit here by claiming that they no longer have the resources to provide on-the-job training. Who does? Where else will new scientists get on-the-job training?

To some extent the demands of business and government have simply shifted a burden onto the academic suppliers. We need to question whether that shifting is correct and appropriate. (Actually, one could take the perspective -- again by changing one's metaphysical point of view — that the system could be serving industry and government very well right now, if only they would recognize it. Although it may be true that new PhDs do not provide what industry and government need, the individuals who have been denied tenure after six years of collaborative work and excellence in communication could do very well in those other environments, and their training has been subsidized by academia already!)

A third assumption that is not universally held at this symposium, but that I believe is more generally held at large, is that the right set of courses can in large measure produce the right product. Furthermore, it is in the contexts of workshops and mentoring, and Joan Garfield's constructive learning, that the interdisciplinary aspects that we have been talking about are most effectively absorbed, confronted, struggled with, and taught. Perhaps the assumption regarding what we do in courses that are very finely structured both in time and space, if not in format, does not provide the best route to achieve interdisciplinary results.

A fourth assumption, which we might prefer to shuffle to the side, is that any student in a statistics program can be molded into the right product. I believe this is simply not correct. There are students who do not have an interest in collaboration; there are students who do not have interest or ability in teaching; there are students who are whizzes at mathematics and are tremendously strong at proving theorems. Perhaps we should not suggest that they should not be PhDs in statistics because they cannot do everything that might be important. Consequently, I feel we have to consider different flavors of statisticians, different tracks, and whether we value them equally.

These assumptions are comforting, but they are not necessary to achieving results. By changing our perspective a little and challenging our assumptions we can sometimes make progress in ways that we had not previously been able to consider.

Joan Garfield summarized the early presentations on client needs by listing three qualities needed in statisticians: they should have teamwork and collaboration skills; they should have communications skills, both oral and written; and they must possess ability to solve real problems. As Duane Meeter commented, these skills, interests, and abilities often describe the complement of those who apply to graduate programs in statistics -- and *that* is a real problem.

Just as journals are at the mercy of those who submit papers to them, so also are statistics programs at the mercy of those who apply to them.

We would like to admit to our programs people who already have an interest in and some initial abilities in these areas, and we would like to make them better. That is what we need, and that is why I am so encouraged by the work of Ed Rothman and of Laurie Snell, which in both cases really reaches out at a very early stage to those people who might otherwise go into psychology or economics or law or business, and captures some fraction of them for the exciting things to be done in statistics.

Moving to another consideration, John Bailar noted, "No one has said, 'I am going to change my life because of what has happened here.'" So let me tell you something about my life, and how what is going on here relates to what I can see myself doing.

Over the next year I will teach three courses and conduct a workshop. Two of the courses are undergraduate courses, one is a course in a professional school, and the workshop in some sense is a graduate course for statistics students.

The autumn course is an undergraduate statistical methods course using Moore and McCabe (1989). As to John's question, "Will it change as a result of what is going on here?," the answer is "yes." I am tremendously taken by Joan Garfield's comments on constructive learning and collaborative involvement in teaching. I have taught this course 15 to 20 times over the past years, and her approach seems to me an answer for which I have been searching. I will try to do something with it.

Is this an interdisciplinary course? In a manner of speaking it is — it uses only real data. It focuses on the real questions as opposed to the statistical questions — that is, one is not done after getting the P-value, and the class often draws on collaborations that have been engaged in with other individuals. But it is also not interdisciplinary; in fact, a section of the course has just been split off for economists so that they will not have to listen to examples drawn from biology and medicine. So although I have changed a little, it is perhaps not enough.

The other courses I teach are more directly relevant to what has been considered in this symposium. Interdisciplinary work is inherently cross-cultural. To work with people in teams requires one to acquire another culture. My winter undergraduate course is on quantitative reasoning. It was designed for humanities majors and is taken largely by people in the humanities, social sciences, and so forth. My purpose in that course is for the students to consider what it means to present an argument, what the nature of evidence is, what constitutes reasoning, and what counts for proof — in mathematics, in statistics, in epidemiology — and to compare those, say, statistical arguments to what constitutes an argument or proof in history, or in literature, or in sociology.

Every discipline has its standards for evidence and its own modes of argument. Those for mathematics are very different from those for statistics, which in turn are somewhat different from those for epidemiology. All these things are worth knowing, at least in comparison to one another, in a liberal-arts education. I have these students, mostly freshmen and sophomores, actually read papers from the *New England Journal of Medicine*, which they find intimidating at first and then kind of fun. I have them read a paper from *Biometrika* -- after a fashion. This is a paper on estimating authors' vocabularies and on deciding authorship. I happen to know this paper fairly well, and in particular I know where all the skeletons are buried. I try to help them understand the structure of the argument in that paper, to understand that an argument is there, that there are rhetorical devices being used, and to help them understand at whom the exposition is being aimed; what the authors want those people to take away when they have finished the paper; what the authors want the readers to assent to; and, to some extent, how and why the authors "swept some things under the rug."

The class reads *The Mismeasure of Man* by Stephen J. Gould (1983), partly for its argument and partly to try to get students to understand that some of the biases discussed in the first half of the book are employed, apparently without recognition, in the second half. But as Jean Thiebaux would have it, this is really a philosophy course focusing on evidence and knowledge. Is it interdisciplinary? It is more cross-cultural, and it is a cross-culturalism that cuts in the other direction; this is cross-culturalism for the non-statisticians in Jon Kettenring's or in Phil Ross's group. I hope that this course will train, or at least excite, some individuals who will do a better job of communicating in those teams with the statisticians on them, and I think it is also important for the discipline of statistics as well.

By implication, training statistics students entails a responsibility for those students to become versed in another culture, a process that takes time and commitment. It is a task that is never completed, and that suggests that perhaps more specialization rather than less, in terms of being interdisciplinary early on, should be the order of the day. If interdisciplinary work is introduced into statistics graduate programs, perhaps individuals should be encouraged to become immersed in just one other disciplinary culture and save the others for later. This immersion involves more, I believe, than merely acquiring interpersonal skills; it requires understanding of how other people argue and how they think.

My spring course is a required course for medical students that they take toward the end of their second year. This course has slowly changed over the last four to five years. Five years ago, fewer than 50% of the students came to more than 20% of the lectures; medical students are very good at voting with their feet. The course has since been renamed, modified, and refocused; it is now called "Epidemiology and Clinical Investigation." The focus has changed from our (statistical) culture to one of their (medical) culture, asking for each lecture, "How will the content of this lecture improve these individuals' ability to practice medicine?" Taking that question seriously changes a lot of what one does, and it seems to have been successful — at least in the sense that now over 90% of the students come to more than 80% of the lectures. The class reads current papers from the medical literature. The emphasis is on critical appraisal of the literature, understanding the strengths and limits of different study designs and providing a framework for what is coming to be called evidence-based medicine — in effect, inference.

These last two courses show, at different levels, that if statisticians take the initiative, very good strides can be made at communicating with other individuals in other disciplines and in other cultures. How do we teach our own statistics students, then, to do this? This is asking how holistic, data-savvy statisticians can be created. The workshop to which I alluded earlier is part of what I consider to be a successful model and has much in common with the Carnegie Mellon approach. That successful model is one based on cross-appointments; but I believe that, while helpful, joint appointments are not enough. John Lehoczky talked about Joel Greenhouse's involvement with psychological statistics at Carnegie Mellon, and Joel has many collaborators; but it is clear that those collaborations spill over onto the students. Through my involvement in the medical school in Chicago, I too have many collaborators that I also "hook up" with students.

How is that done? Merely holding a joint appointment does not do it. Someone in this symposium mentioned the need to have an entrepreneurial spirit. One has to go out and talk to them. When I joined the department of anesthesia, the chairman and I mapped out the high-traffic areas of the department. The highest-traffic place was the so-called coordinator's office where the surgical schedule was made out each day for the next day's surgery. That was a great place because the official schedules were there; when there was a change in Operating Room assignments or someone wanted to know who was in a particular O.R., they came into the coordinator's office. Coordinators used the office for about an hour and a half each day. So we said, "Great! We'll put Thisted there!" So I sit in a tiny room, about half as big as this table, whenever I am in the anesthesia department. It is wonderful because everybody comes by — not to see me, but to check the schedule. And then while they are checking the schedule or looking up something else, they say, "Oh, by the way, I have been working on this problem. Could we get together sometime and talk for a little while?" An appointment is set, and that is when the real work gets done. But in some sense, the real work got done earlier by sitting in that place that is both visible and is clearly part of the department. I have lots of collaborators, and the reason is reflected in what they say when they tell me, "You're the first statistician who talks my language. You really understand what I'm trying to do." Well, in a sense doing that is not too hard, because I always ask them, "What are you trying to do?" And they are very happy to tell me. It is not in me, it is in them, and they see themselves reflected in me, and everybody likes to look at himself. They also say this about my students, though; one of the great quotes goes something like, "I had to make her second author because she contributed so much." That is simply a wonderful thing to say; it makes me beam with pride. The statistics department has lots of students appearing as second and third authors on papers, and it is because we sort of hammer into them that you have to take it on yourself to learn the other culture. If you are working with somebody to distinguish prerenal azotemia from acute tubular necrosis, you had better know what those things are. It is not enough to know only how to pronounce them.

To work on learning the other culture, several years go we started a workshop in biostatistics modeled on the one example I knew then — Bill Brown's at Stanford. Jim Landwehr reminded me that Bill Kruskal also did this years ago at Chicago, and it must be done in many other places. The object is to bring together statistics graduate students, physicians, investigators, clinical scientists, and administrators to talk through problems, to discuss work in progress, and to talk about plans for studies. In the ideal session, a statistics student and a medical investigator who have been working together on a project lead the discussion, with the investigator giving the medical background for the non-physicians and the student discussing the methodological approaches to the questions under study. We have 25 to 30 people who come every Friday afternoon from 3:30 to 5, not generally considered an advantageous time slot.

The results of this workshop have been phenomenal, including many collaborations between students and investigators and several joint publications that do not have me as coauthor but that do have students from the department of statistics. These articles appear in journals such as *Pediatrics*, the *Journal of Neuroscience*, and *Anesthesia and Analgesia*, and our first *New England Journal of Medicine* paper appears this month. These students are mostly from the statistics department. They do not get any course credit for coming to this workshop, nor for working with these investigators. The workshop fulfills no requirements for them of any sort. Our goal is to excite students about the contributions that they can make to real science, and I think it has had some effect in doing that.

What do they get for coming every Friday afternoon from 3:30 to 5? I will tell you my secret: they get cookies. But that is probably not enough. They also get thesis topics, they have gotten jobs, and I believe most importantly, they have gotten tremendous satisfaction from the work they have done.

So to conclude, I believe that a major key to fostering interdisciplinary work is to make it exciting and accessible and real, to do it in the context of a statistics department or any other unit that in some sense is a center of intellectual ferment where exciting and important work is being engaged in. I believe that the group should have as the assumption underlying what it does that interdisciplinary work is not what you do, but rather is really a part of who you are. And I believe that that is what is needed to attract the kind of student who will satisfy the clients and will satisfy themselves. I am convinced that it is worth taking the risks to do that, to make those changes. I am also convinced that "if you build it, they will come."

References

Box, G. E. P. 1976. Science and statistics. *J. Am. Stat. Assoc.* **71**:791-799.

Gould, S. J. 1983. *The Mismeasure of Man*. New York: Norton. 352 pp.

Moore, D. S., and G. P. McCabe. 1989. *Introduction to the Practice of Statistics*. New York: W. H. Freeman.

Discussion

DUANE MEETER: Consider a hypothetical situation in which undergraduate students take a mathematical statistics course and get excited about some theories but, to their shock, suddenly find out on getting a degree and going to graduate school that they are going to have to work with other people and in groups larger than two. This makes me feel that we have to move upstream, look at our mathematical statistics courses at the undergraduate level, and use some of the methods that have been discussed here, for example, ask people to work in teams so that they can expect more of that when they get to graduate school.

Of course, the other problem is that the faculty are not trained this way either, and so it is a tremendous job to change their thinking.

ROTHMAN: I would certainly welcome any suggestions you have about how we can change the faculty so that they think about these things and actually do something about it. Some of them have thought about it. But there are tremendous disincentives to making changes. Frankly, it is far easier to teach a well-defined body of theory. It is neat and fun to simply go into the class and present the standard theorems and be done; you have nicely defined lectures. This interdisciplinary and cooperative stuff is a lot harder. Also, it takes more time to do this right, whereas the other material you can dig out of an old book you had when you were a student.

At the University of Michigan—this is not said out loud, but you know it is true—people do not get promoted for outstanding teaching. Even if your colleagues think very highly of you as a teacher, the basis for promotion at the University of Michigan is ten outside letters. Where are you going to get ten outside letters? One must write a pen sketch of who those people are, and people in the field are contacted who are experts in this specialty and that specialty. They are themselves not universal masters, and they say, "Here are a few of the good papers that I understand, but I do not know enough about those other papers." This results in a situation in which assistant professors have a tremendous disincentive to get engaged in interdisciplinary education. Their best bet is to write research papers and get published in the *Annals of Statistics* and perhaps a few other places. That is not to say that Michigan will not promote other people; I am an example of that, but it is pretty rare. I think that changing that is the biggest challenge.

MORRIS: However, there are two problems. You spoke about measurements in your presentation; one problem I would like you to address is how you measure good teaching, because it is easy in some sense to measure good research; you get those ten letters and see what the person is like. It is harder to observe teaching, and so how can we measure that? I do not expect an answer today, but that is one problem.

When we teach well, looking at this as a collective team process, we are helping hundreds of students to learn; if we could measure how much better off the field is 20 years from now because of what somebody does in his or her teaching, we might find that that person's teaching is more important than some abstract research published in an obscure journal.

ROTHMAN: I think that is the key. I have heard the word "holistic" twice — call it system thinking. The way you measure the performance of anything is in terms of its impact on the aim.

To make it quite simple here, to evaluate teaching, I think it would be wrong to look at things that are too focused on just one aspect of the whole. We are not interested in maximizing individual parts of the function; that is what is meant by total quality. You look at the function of all these variables. And you are right; ultimately, what we want to know is how much learning is taking place 20 years from now. So if we look too carefully at what happens in the classroom, we may lose sight of the real aim.

My attitude about teaching is that it cannot be quantified in such simple terms. I have been the beneficiary of some of those simple measures. I have all the little awards, but that is nonsense, frankly; you can manipulate the audience to get those awards and garner high grades. But what really matters to me are the successes of some of my students and how they did later after having passed my way. That is what I feel best about, not what I did, nor whether this person liked me or not.

I think measuring good teaching will have to be based on some sort of written evaluation. You know who the poorer teachers are, and I think help must be focused on those who are not doing quite the job that ought to be done, on a one-to-one basis. When we set up some measure, people will optimize for that measure as opposed to the purpose of increasing the joy in learning.

ANTONIAK: Having taught for many years, I have to agree. In a parallel field, how do medical schools prepare doctors to recognize that they might possibly make the wrong diagnosis, or make the right diagnosis and provide the wrong treatment, and that in both cases, the patient will die? How do we prepare applied statisticians for dealing with poor data?

ROTHMAN: In medicine, there is no issue of truth to provide a yardstick. They can at best look at the process by which they arrive at some decision, but knowing all the while there are very few gold standards; even autopsies can have errors.

One bad approach that I know they use is multiple-choice tests for some evaluations. It is bad because in a world where women who entered medicine were clearly of a higher intellectual level than the men, thanks to vast levels of past discrimination, women did poorer on that platform. Yet when women were evaluated one-to-one, they did much better than their male counterparts. So some of their measures were simply nonsense.

At many good medical schools they simply throw out all that grade nonsense and report Pass/Fail; everybody of course passes, because if you do not pass, you are out of medical school. That is a much closer approximation to the truth than reporting on some numerical scale. This idea of connection to an outcome, a real outcome, I think is misleading. I wish I had more time to go into this.

MYERS: Are you a voice crying in the wilderness, or does anybody pay attention? What is the attitude of your colleagues at the university?

ROTHMAN: There are a few people in the department who agree. There is another group that is on the fence (you will find this is true everywhere), and there is a group that is in opposition. Our department has calmed down over the years. I do not think I am a voice in the wilderness, but there certainly is not a big parade out there pushing this along. I believe there are people who understand much of what Deming has been saying and are implementing it, but a vast majority are not changing. Whether they believe that the change is right or otherwise, they are simply not changing. I do not see any radical changes taking place.

MYERS: You essentially said in a slightly different way that all of this is clearly not having much success.

ROTHMAN: We have got to change ourselves. One could sit back and say, "All of this and a buck and a half will get you a cup of coffee." What I am suggesting is that at least those of you who are here can change. It is such a simple idea that if you change yourself, that is all you can do, really; you cannot get other people to change, but if you yourself change, we are going to make something happen.

We have to increase what we consider to be our faculty, and include adjunct faculty. We have this tremendous industry resource. We choose to view statisticians in industry as not academic statisticians; that is ridiculous! There is much more that industrial and academic statisticians have in common than what they do not, and we ought to be bringing people from industry into the classroom instead of only sending our classroom theoreticians to industry. It is a two-way street.

There are a lot of little changes we can make. It is a gradual process. Stephen Fienberg said we should measure progress on the scale of generations. I do not know about that, but I feel that there is something of a moral obligation that when you have knowledge about how things ought to be, you should not sit idly by and let other people define the world for you. You need to say, "If that is what I think is right, I am going to make some changes along those lines."

The fact that we have a big barrier and a lot of people who are opposed to changing or doing anything differently: so what? Look at the fellow who proved Fermat's last theorem. He went seven years without a publication. I wonder if he talked about his paycheck at the end of each year the way we might, when he did not get any salary increase. He was afraid to tell people he was working on it. He had one colleague in the department review his work, apparently.

If you think what we say here is right, make some changes.

SOLOMON: I declined John Tucker's earlier invitation to talk about the statistics department at North Carolina State University, but instead I am going to do a little marketing for the *Journal of Statistics Education*. It has been referred to a couple of times in these two days. The first issue of that journal, which is accessible electronically in a number of ways, appeared on July 19th. The founders had two motivations in mind for establishing the journal that are relevant to this symposium. The primary one was what I hope is the obvious one, and that is to provide a vehicle for exchange of information on postsecondary statistics education. But there was a secondary goal that relates to the reward structure that we have talked about and has come up a couple of times in the course of this symposium. That goal is to provide a rigorously refereed, and, hopefully, therefore respected outlet for scholarship in statistics education, one that will provide some tangible fodder for promotion and tenure files. Whether it achieves that goal depends on whether all of you begin to submit papers to it. Part of the excitement of it is that it is distributed electronically, and so offers opportunities that simply are not available in the print media. For example, an article in the first issue written by George Cobb describes the results of a sequence of NSF-sponsored projects in statistics education. The article itself is a summary of those projects, but if you wish, you can request from the *Journal of Statistics Education* additional information at various levels that will provide, for example, the final report of the project team or, in fact, some of the data sets or work sheets that were prepared in connection with some of these projects. That is part of the vision for this journal, to make materials as well as articles available electronically.

There are some brochures outside at the registration table, and there is going to be a poster session at the Joint Statistical Meetings in which the *Journal of Statistics Education* managing editor and the editor will actually demonstrate with a computer how to access the journal and let you see what it looks like on-line. I would be happy to talk privately to anybody who is interested in more information.

PREM GOEL: A question for Laurie Snell: How do you assess students in a course such as Chance?

SNELL: We benefit from the fact that Dartmouth, like many other schools, has wonderful grade inflation; for example, the average grade at Dartmouth is more than B+. Thus we are talking about giving grades of A, A-, B, and B+, and so we do not have to worry about huge distinctions. The actual final grade is based on the following components.

The students keep a journal of what has taken place in the course. The journal is used for students' comments about the articles that they were a little shy to make in the class or in their discussions. They put in the journal solutions to problems, if they have worked out of Friedman, Pisani, Purves and Adhikari, and so on. The students do all of the review exercises in the relevant chapters, which is usually about two-thirds of the book, pretty much on their own. Very little time is spent discussing that. They also hand in a lot of homework. At the end of the course they do a final project that is related to something they were interested in that came up in the course. That project is a substantial effort. Students do a poster show, and there is a kind of casino that goes with it; that, too, is a big event, and students seem to like it very much. A student's final grade is our subjective assessment of the journal, the homework, and the project. With there not being a lot of grade range to worry about, the true answer to your question is that assessment is not taken very seriously.

WARD: Is this class one large group?

SNELL: The schools where I have been involved with teaching the class have by chance been three different ones: Princeton, Dartmouth, and the University of California, San Diego. Two classes were of size 60, and one was 30. Sixty is pretty large, although I personally always team teach it with somebody else; I would not teach it any other way. There are simply too many ideas floating around at one time for one person to handle, and it is very convenient to be able to say, "Well, I do not know anything about that but maybe Joe does." The students love this, of course; there is nothing they like better than to see experts arguing. I do not know why they love that so much, but they do.

JOHN McKENZIE: How do you gather the information? How do you get the clippings?

SNELL: I am exposing all my secrets and am probably going to go to jail even faster! I scan through Mead Data Central, Inc.'s Lexis and Nexis for articles on the topics that have been interesting in the past, and then I try to find new ones by searching on such topics as probability theory and statistical theory. I thus get a long list of candidate topics. One of the wonders of modern electronic roads is that all these newspapers—the *Los Angeles Times*, the *New York Times*, yesterday's paper—are all there today, and so you are at most one day late in searching for these things. When I find articles that are interesting, I try to write a little abstract of them. Then I simply download these articles to our database. All of this is done electronically; I never print anything.

MYERS: How are those newspapers accessible?

SNELL: Via gopher. Sign on to chance.dartmouth.edu, look into the Pull Text folder, and there they are.

MYERS: But prior to that?

SNELL: They are on Lexis, Nexis, or any of a number of databases that have the current major newspapers on them. Finding articles from journals is not as easy. Newspapers seem to be more casual about allowing their current issues to appear on the database. The journals do not like that, and so they tend to be a month or so behind.

THISTED: How much of your time is spent preparing for and teaching this course?

SNELL: I guess all, most, lots. But then it is just something of a hobby with me; it is in a way my final fling. Other people have taught it; for example, Tom Moore has taught it at Grinnell and Bill Peterson has taught it at Middlebury, and they have followed a somewhat less crazy route by picking three or four topics, such as quality control or streaks in sports, so that to some extent they can prepare that material ahead of time. On the other hand, their courses were actually part of a writing curriculum course that, they said, was really murder; to try to do two things like this at once is just astronomically difficult. There is no doubt that it is a difficult course to teach.

When we applied to renew our NSF grant, they made it perfectly clear to us that if we would figure out some way to allow somebody in, say, Oshkosh to know what lesson one is, lesson two, lesson three, lesson four, they would like this course much better. They are very nervous, obviously, about how likely it is that this Chance course will make an impact in the sense of people simply being able to do it, just having the time and effort. We hope people will do something much more modest, namely, in their own courses or just occasionally, use one or two of these things that they get from our CHANCE database. I am sure many people already do it; the idea of using current events is certainly not something we invented. I suspect many of you talk about something you read in the *New York Times* and so on, but this database just makes it a little more accessible.

MYERS: A question for Prem Goel: To what extent does computing come in to play in the Ohio State course on statistical factors?

GOEL: Throughout the year computing is an integral part of the course. You cannot do statistics today without computing.

MOHAMED MADI: What happened to the more traditional courses in the OSU statistics program?

GOEL: In the first year there are three sequences under way. The first is the mathematical statistics sequence, which is basically an introductory course. The second sequence is the real analysis sequence for the whole year, and the third sequence is the statistics practice course I described. Basically, everybody takes all three courses each quarter; note that the three books they have been using will change next year.

KETTENRING: Just to comment on the reaction from the students to the unstructured course, I have two thoughts. One is that their reaction may in some sense be an early litmus test, identifying students who are going to find these sorts of statistics that we are talking about beneficial. My second thought is that we are really providing these people a wonderful, early exposure to what life is like in the real world, where it has nothing of that structure, and I hope that they appreciate some of what you are doing.

ROTHMAN: Students find identifying a series of questions to be less satisfactory than seeking specific answers. However, concentrating on what are the right questions is a better model than what you may see in talking with the students. It is not always acceptable to the student driven by a course grade, and I think that removing that course grade is the key to any successful transformation.

WARD: Again, what about the course grades for the OSU course, since it has been brought up again. Does Prem Goel have any comments about grades?

GOEL: For the statistical practice course?

WARD: Yes.

GOEL: The grade was only Satisfactory or Unsatisfactory. Competition is a problem in grading. If you have four people working together, you cannot have grades of A, B, C, D, and F, because then you will have problems in terms of their cooperation with each other.

WARD: Do the students have some idea of the course objectives, so that they can know how grades of Satisfactory and Unsatisfactory are determined?

GOEL: I believe they do, but still, it is something they have not seen in their undergraduate years, and so they are uneasy about it in the beginning.

TUCKER: Are the faculty for the statistical practice course members with secure positions who do not have to worry about how devoting time to such a course might affect a tenure review?

GOEL: Yes. Both of the faculty members involved are senior associate professors, and so they are not worried about such things. We had four volunteers for this course, and I, as department chair at that time, chose two of them knowing that they would work very well together. It is important that a good team be selected and that the team people can work with each other. In this case, both these fellows worked with each other quite well.

ROSENBERGER: This is a first-year course?

GOEL: That is correct.

ROSENBERGER: We at Pennsylvania State University have worked on something similar in our first-year courses, but the question we discussed and have not resolved is whether there should be an advanced, second-year data analysis course for PhD students. It would replace a menu of topic courses that they now take; do you think from the experience at OSU that such a thing should be attempted?

GOEL: That is an interesting question, because we were not sure ourselves whether it should be a first-year course or a second-year course. However, if you want to implement something you have to keep in mind what the context is going to be. Quite a few of the OSU statistics faculty were very unhappy about so-called tool box courses that were being taught in our program. We considered this to be one way to solve that; it would be introduced the first year, and, hopefully, after three years of evaluation, it would then be decided if it should be in the second year. However, by that time, the tool box courses will not be in the program anyway. So it is not clear that the first year is the best time to do it; in that I agree with you.

SOLOMON: We at North Carolina State University have added the requirement; we have had for some time a second course in methods. It was to address what our graduates, who in our case are often employed in the pharmaceutical industry and similar places, were telling us they were using in their jobs. To try to keep the curriculum manageable, we decided to collect many of those topics into a single one-semester course that introduces students to the existence of the topics, points them to the literature, and gives them a notion about the topics without going into the kind of detail given in, for instance, a whole course on survival analysis. This sort of survey is what we at NCSU call intermediate methods.

GOEL: I forgot to mention that we at OSU use seminars partially for the purpose of educating the students. We have nearly 2 seminars per week; in a 10-week quarter we may have about 17 seminars. One is supposed to be theory and one is supposed to be applied, again a counterpoint. The seminar is a course for students; they get one credit hour for it every quarter. Students are to attend 9 out of 17 or 18 seminars per quarter. We invite a number of faculty on campus for either of the two seminars each week so that they can present what problems are being addressed at Ohio State. It becomes a good tool for students to learn about the scope of statistics. The only problem is that students do not necessarily understand all the details, and so we tell them, "If you can understand the first ten minutes, you are doing fine; do not worry about it."

SOLOMON: We at NCSU are large enough that we can afford to offer quite a variety of courses, and at the graduate level we have introduced over the years a number of courses called Applied X, for example, applied time series, applied whatever, with the target audience being graduate students from other disciplines. But who is taking those courses? All the statistics majors, the master's degree students, as well as the originally intended audience. But they get the word that it is useful stuff.

ROTHMAN: Concerning the statistical practice course's content, I noted that regression and all the standard things were included, but one thing I did not see, which may nevertheless be in the course, is a requirement that the student give a presentation. We talked about written expression. But I think that what we have now is a graphics medium that enables us to express ourselves in a very different way—much the same way as we present results in a scientific paper. Presenting the results of an analysis in a graphics mode has only begun to get the attention of our faculty and researchers but should, I think, be integral to such a course. Do you do that at OSU?

GOEL: Not at present, because that is more or less covered in the consulting course rather than in the statistical practice course.

ROTHMAN: It would be nice to hear what in fact you do, what principles you advocate and so on. Another aspect of the course has to do with design. Design often focuses on randomization and procedures like that, when in fact most of what is out there are observational studies. Yet I believe we have a lot to share about what are interesting things to look at in observational studies, such as what some of their limitations are. To what extent do you talk about that?

GOEL: That is addressed quite extensively in the course, in the first quarter, in fact.

MYERS: To what extent do you include things in the course that require students to look at the literature?

GOEL: Not to a great extent as yet. If we ever make this course a second-year course, we may do more of that. They currently look for some examples of real data that relate to projects, but they do not look at the literature as such.

MYERS: I teach a spatial statistics course in the mathematics department that is populated almost entirely by students from the statistics department. I require that they do three journal-article reviews. These are probably not done in real depth. I ask them to identify what problem is discussed in the paper, use some kind of "legal review" technique or approach, and perhaps make some comments about the usefulness of what has been obtained. In part, I really want them to simply look at the literature and see what is there, but it forces them to use the literature.

GOEL: I do quite a bit of that in the second core consulting course.

FIENBERG: For a similar course at CMU, one that evolved away from what I did some ten years ago rather than solidified it, efforts are now made to reach out and draw in both faculty members and graduate students from other areas. The very exciting thing that happened to our curriculum and our students over the last six years was that we began involving students early on in projects. Even master's degree students have actually either collaborated with graduate students or been part of a team, albeit sometimes a junior member in a team. Currently, I am at the point where they have written up the material and have presented it, which is sometimes done jointly with a graduate student in another department. You often get a four-person team—two faculty members and two graduate students.

I believe you have to actively reach out; again, that is an even more complex kind of mentorship-apprenticeship relationship, but it is absolutely essential in order to do what I call interdisciplinary research. You must actually immerse yourself, even for a restricted period of time, in that culture. You have to go to them; you cannot expect them to come to you, because when they come to you, they will narrow the question that they ask and you may not know what they are really doing.

MYERS: A question for Steve Fienberg: Have you thought about the distinction between statistics in the law and statistics in public policy?

FIENBERG: What I call statistics in the law is a very focused and structured thing. Almost everything else I do is in a much larger sense statistics and public policy. Laurie Snell's presentation was very illustrative. There are a number of absolutely critical overlays that we do not teach our students in a typical statistics class, and I think that they are essential. The word "ethics" does not come up very much in the statistics curriculum. But if you go out and do a real project, it comes up all the time in disguise.

The notion of social values defining how fields structure their language and their investigations, and how public policy making is structured is something that nobody talks about in statistics journals and professional meetings, let alone with statistics students. When you get into things such as this, you are drawing in a much more complex array of issues than those that we traditionally talk about within statistics—ones that would come up much more naturally in, perhaps, an ethics course in a philosophy department. I think that that is an exciting aspect for me as a statistician, and it raises yet another set of interdisciplinary questions that I find quite challenging.

BAILAR: I said yesterday that I had set aside the entirety of what I came prepared to talk about and presented something entirely different. One of the things I left out has to do with the teaching of ethics. I teach a course in advanced epidemiology. This is presented is a combined department, so that there is also a lot of biostatistics in it. It is basically designed to be the last course that our PhD candidates take before they spend full time on their dissertations.

I started with one hour — by student demand, it is now up to six, and next year it is going to go up to 10 — in which we talk about ethical problems. It has something of a seminar format; the students do as much talking as I do, sometimes more. We talk about the smaller ethical issues; we do not spend much time on fabrication of data, falsification, and plagiarism — the big three of scientific fraud. We do not spend any time on those because everybody agrees those are wrong! There are no issues with them.

We spend a lot of time talking about cutting corners, which I feel is a much more serious threat to the progress of science, and about how to identify the lesser sins: mainly, not telling people about all the things that went wrong in your experiments, or about how much you are leaving out that might reflect badly on the initial design, or when in the course of the work you the statistician or you the investigator actually came up with the hypothesis that you present with a P-value. All of this I present in an ethical forum — stressing what actions are wrong — and the students gobble it up. Every year they have asked for more, and they have said something I think is a terrible reflection on the prior teaching they have had both in our department and, I suspect, everywhere else. They say that nobody has ever mentioned these things to them anywhere before as ethical problems. It makes my colleagues very uncomfortable, but the students are clearly very interested in this.

THISTED: I would like to ask John Bailar if his colleagues are uncomfortable about making value judgments or if they are uncomfortable in talking about it to the students?

BAILAR: Half of it, I believe, is that they are uncomfortable talking about this. They do not like to think that deliberate deception, cutting corners, and shaving results is really a part of science; yet you know and I know that it is, and we had better recognize it so that we can do something about it. The other half is that they do not feel this is really part of statistics, and I think that they are flat wrong.

MORRIS: Regarding on-the-job training, we need to prepare the students essentially to be fast learners or to know where the references are. If we are not going to teach them everything, we need to give them the ability to teach themselves and, obviously, to network with other statisticians.

ROTHMAN: I think ethics is an integral part of what we do as statisticians. I introduce it in the Michigan class as one of the many constraints in designing a study. I think it is interesting for students not just because it is important, but also because of the relatively recent history of ethics in experimentation. We have names for some of the things that you talked about, going back, for example, to Charles Babbage, such as "cooking" data and trimming data, where cooking means serving up only what you feel is fit for consumption to support your case. This is not something that has been around for centuries or even a hundred years. Much of the constraint randomization work came out of the Nuremberg conferences and was only done in 1948. I think ethics is something to truly get students thinking, and that it is absolutely essential for them to grasp.

FIENBERG: When I first raised this ethics dimension, I did so in a very special context, and I think it is great that John Bailar brought out some of these other aspects that Ed has amplified. It is striking to me that over the last decade, I found myself drawn more and more into issues to which the word "ethics" could be attached that did not start out looking to be that. A number of years ago I was involved in a project that ultimately came under the label "sharing of research data," where the goal was to achieve access to data sets largely for secondary analysis, something that fits quite naturally into the theme of this symposium because it provides terrific resources for classroom use. We discovered the problems about access; some of them are ethical and some of them are practical. In going from field to field, you discover that within fields the ethical problems are rarely discussed. As that work has evolved, so also has my involvement with it. It seems evermore the responsibility of statisticians, not because we are a higher moral authority, but because of our ability to generalize, pulling together the examples that we see in multiple fields, and pointing out ethical issues that others have not always identified as such.

I take a number of John Bailar's examples as illustrative of that. I see it everywhere in what I do and I now view things differently than I did earlier. I now look for vehicles for sharing that with students that I did not use a decade ago, before I became involved with this.

BAILAR: We at McGill spend an hour, sometimes two hours, talking about sharing data, during which I always cite the National Research Council's nice report *Sharing Research Data* (NRC, 1985).

JAIME GRABINSKY: Many times when we are teaching civil engineers, mechanical engineers, or physicists, we do not present the problems that really are of concern to them. For example, I know of one book on engineering statistics that has many interesting problems for civil engineers but was of course written by a civil engineer. But I am not able to find the types of books and materials that deal with other subjects, especially for engineers. I have looked through many books in physics and many books in statistics, and I know there are many important statistical problems in physics that are not appearing in the statistics books as well as many topics in statistics that are not appearing in physics books. So there is a lack of communication, a lack of mental infrastructure, if you will, that we need to achieve.

SOLOMON: I came across a book that is used in a graduate course in the chemistry department, *Statistics for Analytical Chemistry* (Miller and Miller, 1993). It is not very thick, yet in a couple of hundred pages it covers what, if we were teaching the material in our statistics department, would require chemistry students to take four or five courses; they do it all in a couple of hundred pages in the one course in chemistry. Also, in briefly looking through the book, I found that it seems to be largely what we statisticians would judge as correct and even uses much of our language.

SNELL: This relates to a problem that has not been solved and I think never will be solved. In a way, the most depressing thing I heard at this symposium was that a person would have to spend a year or two to learn what "bootstrap" was all about. That was one thing I learned a few years ago and thought was a wonderful idea, and I thought I understood it. I really do not know what the answer is to the question of how to teach it all. I do not know much about statistics—it is not my field; but in mathematics people have talked about the same problem. There everybody says, "What we need are students who really can think on their feet and are not interested in all these specialized things." They talk about horizontal knowledge instead of vertical knowledge—William Thurston is a key spokesman for that. Yet I know very well that when such a student applies to Princeton for graduate work in mathematics, they want to know whether the student knows this or that and so on; it is probably true in statistics, too. In other words, when you admit students to graduate work in statistics, do you really look at whether they are interested in different ideas and different kinds of work, or do you find out what courses they have had and how much statistics they know? Is horizontal knowledge valued, or is the focus on how the student is going to be able to handle this very difficult program and the rather theoretical work he or she faces? And there you probably prefer the applicant who makes you think it is going to be a safe bet.

I do all of my current teaching with somebody who is now regarded as so brilliant that everybody would give him tenure, no problem; but he had a lot of trouble getting into graduate school because he really was interested in ideas and not specific facts.

References

Miller, J. C., and J. N. Miller. 1993. *Statistics for Analytical Chemistry.* 3rd ed. Englewood Cliffs, N.J.: Prentice Hall. 256 pp.

National Research Council. 1985. *Sharing Research Data.* Committee on National Statistics. Washington, D.C.: National Academy Press.

Appendices

Appendix A

Speakers and Session Chairs

Jerome Sacks is director of the National Institute of Statistical Sciences, a member of the Executive Committee of the Board on Mathematical Sciences, a fellow of both the American Statistical Association (ASA) and the Institute of Mathematical Statistics (IMS), and a member of the ASA Office of Scientific and Public Affairs Advisory Committee and Committee on Professional Ethics. He has worked collaboratively in high-technology areas with engineers, computer scientists, chemists, and others in the industrial and scientific community. He has been professor of Statistics and head of the Statistics Department (whose formation he led) at the University of Illinois at Urbana-Champaign, program director for Statistics and Probability at the National Science Foundation, and a member of the Committee on Applied and Theoretical Statistics. He co-chaired the IMS panel that produced the 1988 report *Cross Disciplinary Research in the Statistical Sciences*.

Jon R. Kettenring is executive director of the Statistics and Economics Research Division of Bellcore, chair of the Committee on Applied and Theoretical Statistics, a fellow of the ASA and American Association for the Advancement of Science (AAAS), and a member of the IMS and International Statistical Institute (ISI). He is on the board of directors of the Interface Foundation and on the board of trustees of the National Institute of Statistical Sciences, is chair of the Management Committee for the *Journal on Computational and Graphical Statistics*, is former chair of the IMS Subcommittee on Cross-Disciplinary Research, and has been vice president of the ASA. His interests include communications between statisticians and engineers/physical scientists, the use of graphical displays, exploratory data analysis, and methods of cluster analysis.

Peter J. Bickel is professor and Statistics Department chairman and former dean of Physical Sciences at the University of California-Berkeley. He is a member of the National Research Council's Commission on Physical Sciences, Mathematics, and Applications, the Virtual Commission on Forensic Use of DNA Typing, and a former member of the Board on Mathematical Sciences. He is a member of the National Academy of Sciences and the American Academy of Arts and Sciences, a fellow of the ASA, a fellow and former president of the IMS, a fellow of the AAAS, president of the Bernoulli Society, and a member of the Royal Statistical Society, the ISI, and Sigma Xi. He has received the Committee of Presidents of Statistical Societies Award and has been a Wald Memorial Lecturer, a MacArthur Foundation Fellow, a NATO Senior Science Fellow, and a Guggenheim Foundation Fellow. He has served as a consultant to a number of organizations, including the University of California-Berkeley Graduate Division, the Environmental Protection Agency (EPA), the State of California's Department of Human Services, the Nevada State Gaming Control Board, and the Institute for Energy and Environmental Research, and has also served as a member of the NRC's Task Group of the Panel on Amplitudes of Coastal Surges for Hurricanes, the Building Research

Advisory Board's Panel on Mapping Mudslide Insurance Hazard, and the Panel on Applied Mathematics Research Alternatives for the Navy. He currently serves on the board of trustees of the National Institute for Statistical Sciences.

N. Phillip Ross is director of the Environmental Protection Agency's Environmental Statistics and Information Division. He also holds an adjunct appointment with the Department of Mathematics and Statistics at American University. He is an active member of the ASA and past chair of the Section on Environment and Statistics. He represents EPA interests on a number of national and international organizations. He is responsible for EPA statistical policy and is currently directing EPA efforts toward the establishment of a Federal Bureau of Environmental Statistics.

John C. Bailar III is acting director of the Department of Epidemiology and professor of Epidemiology and Biostatistics at McGill University. He is also scholar in residence with the NRC Board on Environmental Studies and Toxicology, and medical scientist with the Division of Clinical Epidemiology at Montreal General Hospital. He is a member of the Institute of Medicine, a fellow of the ASA and AAAS, a member of the Biometric Society and ISI, a member and former president of the Council of Biology Editors, and a member of the Society for Risk Analysis. He is currently a MacArthur Fellow, a member of the *New England Journal of Medicine* Editorial Board, and the ASA Environmental Statistics Technical Advisory Committee, and has been editor-in-chief of the *Journal of the National Cancer Institute*. He is a member of the Board of Governors for the Research Triangle Institute and the Board of Trustees of the National Institute of Statistical Science. His research interests include cancer epidemiology, randomized clinical trials, risk assessment and environmental epidemiology, and the development of related statistical methods and theory.

James M. Landwehr is supervisor in the Statistical Models and Methods Research Department at AT&T Bell Laboratories, Murray Hill, N.J. He is a fellow of the ASA and AAAS and a member of the IMS, the Mathematical Association of America, and the National Council of Teachers of Mathematics. He is chair of ASA's Section on Statistical Education, and council representative from ASA's Section on Statistical Graphics. He served as member and vice-chair of the ASA-NCTM Joint Committee on the Curriculum in Statistics and Probability, which led to his role as co-principal investigator in the Quantitative Literacy Project for introducing statistics and probability into middle schools and high schools, and he co-authored two of the QL books. Currently he is a member of the ASA-MAA Joint Committee on Undergraduate Statistics, and he served on the NRC Panel on Non-Standard Mixtures of Distributions. His interests include statistical applications and collaborations, and research on applied statistics methodologies.

H. Jean Thiebaux is a program director at the National Science Foundation in the Division of Mathematical Sciences. She has taught statistics and probability, at both graduate and undergraduate levels, since 1964 at universities in the United States and Canada. Her teaching and research, as well as her own academic preparation, have been largely interdisciplinary with extensive experience and collaboration in the sciences. She was part of the biostatistics program at Stanford University while she was working for her PhD. She designed an innovative master's program in biostatistics and epidemiology at the University of Massachusetts-Amherst. Following two years as consultant and visiting scientist at the National Center for Atmospheric Research, she spent 10 years on the faculty of the Department of Mathematics, Statistics and Computing Science at Dalhousie University in Canada. She was a full professor and coordinator for an interdepartmental program in atmospheric science prior to her return to the United States to work for NOAA. She is the author of two books on statistical theory and methods, specifically written for use in the ocean and atmospheric sciences. She is a fellow of the Royal Meteorological Society and a member of the ASA and the International Environmetrics Society. Her research interests include statistical modeling and estimation for spatially coherent systems.

John Lehoczky is professor and head of the Department of Statistics at Carnegie Mellon University (CMU). He also serves as Program Coordinator for the department's NIMH Training Program in Psychiatric Statistics and is involved with the CMU Statistical Center for Quality Improvement. He served on the BMS Committee on the Mathematical Sciences in High-Performance Computing and Communication. He is associate editor for the *Journal of Real-Time Systems*, a fellow of both the ASA and IMS, and a member of ISI, AAAS, IEEE, ACM, The Institute of Management Sciences, and the Operations Research Society of America. His research interests include stochastic processes and their application to computer and communication systems, real-time computer systems, mathematical finance, psychiatric statistics, and industrial statistics.

Joan B. Garfield is associate professor of mathematics and statistics in the General College at the University of Minnesota, where she served as coordinator of research and evaluation from 1982 to 1986. She is a member of the ASA, the International Association for Statistical Education, the Mathematical Association of America, and the American Educational Research Association. She is secretary and newsletter editor for the International Study Group for Research on Learning Probability and Statistics, a member of the Joint Committee on Statistics of the ASA and MAA, and a former member of that committee's Statistics Focus Group. Her research is focused on the teaching and learning of statistics and probability, evaluating the effectiveness of different instructional approaches, and developing assessment instruments to evaluate statistical learning. She writes a regular column describing current research on teaching statistics for the international journal *Teaching Statistics*, and coauthors (with Laurie Snell) a column on resources for teachers in the new *Journal of Statistics Education*. She is involved in several NSF-funded statistics education projects.

Carl N. Morris is professor of statistics and professor of medicine at Harvard University. He is a fellow of the ASA, the IMS, and the Royal Statistical Society, and a member of the Biometric Society and the ISI. He has been a member of the Committee on Applied and Theoretical Statistics and the CATS Panel on the Combination of Information, the Board on Physics and Astronomy's Committee on Army Basic Scientific Research, and the NRC Panel on Military Manpower Forecasts for Small Areas. He has served on the ASA Board of Directors and as an IMS Council Member, was executive editor for *Statistical Science* and also theory and methods editor for the *Journal of the American Statistical Association*. He worked on many applied projects at the Rand Corporation, including the Health Insurance Experiment and two other national public policy experiments. He was formerly professor and director of the Center for Statistical Sciences at the University of Texas at Austin. His current interests include hierarchical models, empirical Bayes methods, exponential families, and a wide range of statistical applications, including health policy and sports.

S. Rao Jammalamadaka is professor and chairman of the Department of Statistics and Applied Probability at the University of California-Santa Barbara. He is a fellow of the IMS, the Royal Statistical Society, the Institute of Combinatorics and Applications, and the ASA and is a member of the ISI, the Indian Society for Probability and Statistics, and the Mathematical Association of America. He has held visiting positions at the University of Leeds (United Kingdom), Uppsala University (Sweden), the Indian Statistical Institute and the Indian Institute of Technology, the Chalmers Institute of Technology and Göteborg University (Sweden), and the Australian National University. He is a member of the ASA Committee on International Relations in Statistics and the editorial board for *Statistics and Probability Letters* and the *Journal for Nonparametric Statistics*. His interests include nonparametric statistical inference, limit distribution theory and asymptotic efficiencies of test procedures, directional data analysis, goodness of fit tests, and improving the university training of statisticians.

Daniel L. Solomon is professor and, since 1981, head of the Department of Statistics at North Carolina State University. Previously he was professor of biological statistics at Cornell University. He is a fellow of the ASA, and a member of the IMS, the Biometric Society, and the ISI. From 1985 to 1989 he served as editor of *Biometrics*. He is currently a member of the NRC Committee on National Statistics, and the Panel for Computing and Applied Mathematics, vice chair of the Board of Trustees of the National Institute of Statistical Sciences, and president of the Southern Regional Council on Statistics. He is a member of the ASA Committee on Publications and Committee on Women in Statistics, chairs an ASA task force on new communications technologies, and is the ASA representative to the COPSS Presidents' Award Committee. The NCSU Department of Statistics, which has developed modern instructional and research computing and communications facilities and has been integrating them into its curricula, spearheaded the establishment of the new electronic *Journal of Statistics Education*. Its graduate programs have long required cross-disciplinary experience, and co-majors may be elected with, say, genetics or economics.

Edward D. Rothman is professor and director of the Center for Statistical Consultation and Research at the University of Michigan-Ann Arbor. He is former chair of the Department of Statistics. He is a member of the ASA and the IMS. He has publications in statistics, genetics, and engineering journals. Much of his current work is in the area of quality. His work in this area has been primarily with W. Edwards Deming.

J. Laurie Snell is professor of mathematics at Dartmouth College. He is a fellow of the IMS and a member of the American Mathematical Society and the Mathematical Association of America. He was a Fine Instructor at Princeton University before joining Dartmouth. His teaching and research have been mainly in probability theory. He started his teaching career working with John G. Kemeny and Gerald Thompson to develop the course Finite Mathematics for students outside the sciences. He is finishing his teaching career working on the development of a course called Chance as another introductory course for students in all fields.

Prem K. Goel is professor and former chairman of the Department of Statistics at Ohio State University. He has been a member of the Committee on Applied and Theoretical Statistics and CATS' Panel on Combination of Information, a program director for the Statistics and Probability program at the National Science Foundation, and an associate editor for the *Journal of the American Statistical Association*. He is a fellow of the ASA, the IMS, the Royal Statistical Society, and AAAS. His interests include Bayesian decision analysis, probability modeling and statistical inference for applied problems in engineering and social sciences, and linear models and time-series analysis. During the last five years, he has taken an active part in discussing statistics graduate curriculum issues and is especially concerned with teaching statistics in a synthetic manner, rather than as a collection of techniques. Some of these ideas have been implemented at Ohio State University.

Stephen E. Fienberg is Maurice Falk Professor of Statistics and Social Science at Carnegie Mellon University. He was recently vice president for academic affairs and professor of statistics and law at York University. He has been chairman of the NRC Committee on National Statistics, and chaired its Subcommittee on Data Sharing, the Panel on Statistical Assessments as Evidence in the Courts, and the Panel to Review Evaluation Studies of Bilingual Education. He also served on a number of other NRC committees and panels, including the Panel on Non-Standard Mixtures of Distributions, the Panel on Decennial Census Methodology, and the Committee on DOE Radiation Epidemiological Research Programs. He is currently a member of the Panel on Census Requirements in the Year 2000 and Beyond. He is a fellow and former vice president of the ASA, and a fellow of the IMS, the Royal Statistical Society, and AAAS. He is a member of the Biometric Society, the Psychometric Society, the Statistical Society of Canada, and the ISI and is on the board of directors of the National Institute of Statistical Sciences, the IMS Council, and the ISI Council. He has been coordinating and applications editor for the *Journal of the American Statistical Association* and a founding co-editor of *Chance*. His interests include analysis of cross-classified data, statistical inference, federal statistics, statistics and the law, and cognitive aspects of survey design.

Ronald A. Thisted is professor in the Departments of Statistics and of Anesthesia and Critical Care at the University of Chicago. He is a fellow of the ASA and the American Association for the Advancement of Science, and is also a member of the IMS, the Biometric Society, the Association of Computing Machinery, the Association for Health Services Research, and the Society for Clinical Trials. He has been an associate editor of the *Journal of the American Statistical Association*, the *SIAM Journal of Scientific and Statistical Computing*, and the *ACM Transactions on Mathematical Software*. He is currently associate editor (CD-ROM) for the *Current Index to Statistics* Extended Database. In 1981, the University of Chicago awarded him the Quantrell Award for excellence in undergraduate teaching. His interests include statistical computation, statistical education, biostatistics, clinical trials, and meta-analysis.

Appendix B

Symposium Participants

Charles E. Antoniak
Lockheed Aircraft Service Company

Stanley P. Azen
University of Southern California

Jenny A. Baglivo
Boston College

John C. Bailar III
McGill University

Dick Beckman
Los Alamos National Laboratory

Peter Bickel
University of California at Berkeley

Mary Ellen Bock
Purdue University

Jack R. Borsting
University of Southern California

Lawrence D. Brown
Cornell University

Lai K. Chan
University of Manitoba

Clifford C. Clogg
Pennsylvania State University

Keith Crank
National Science Foundation

William Eddy
Carnegie Mellon University

Stephen E. Fienberg
Carnegie Mellon University

Lynn Friedman
University of Minnesota

Bodapati V. R. Gandhi
University of Puerto Rico

Sandra Gardner
University of Toronto

Joan B. Garfield
University of Minnesota

Prem Goel
Ohio State University

Colin Goodall
Pennsylvania State University

Jaime Grabinsky
Unidad Azcapotzalco

Mary W. Gray
American University

Marjorie G. Hahn
Tufts University

Larry Haugh
University of Vermont

Sam Hawala
University of California at Davis

Alan J. Izenman
National Science Foundation

S. Rao Jammalamadaka
University of California at Santa Barbara

Albyn C. Jones
Reed College

Jon Kettenring
Bell Communications Research

Terry King
Northwest Missouri State University

John Kinney
Rose Hulman Institute of Technology

Tomasz J. Kozubowski
University of Tennessee

Frederic C. Lam
University of Otago

James M. Landwehr
AT&T Bell Laboratories

Eun Sul Lee
University of Texas at Houston

John P. Lehoczky
Carnegie Mellon University

Hao Li
University of Arizona

Edward J. Lusk
University of Pennsylvania

Mohamed T. Madi
UAE University

David Madigan
University of Washington

Helene Massam
York University

John D. McKenzie, Jr.
Babson College

Duane Meeter
Florida State University

Chand Midha
University of Akron

Carl N. Morris
Harvard University

Donald E. Myers
University of Arizona

Joyce Niland
City of Hope National Medical Center

Ruth E. O'Brien
National Research Council

Ganapati P. Patil
Pennsylvania State University

Daryl Pregibon
AT&T Bell Laboratories

Darol K. Redmond
University of Pittsburgh

James L. Rosenberger
Pennsylvania State University

N. Phillip Ross
Environmental Protection Agency

Edward Rothman
University of Michigan

Jerome Sacks
National Institute of Statistical Sciences

Stephen M. Samuels
Purdue University

Robert H. Shumway
University of California at Davis

J. Laurie Snell
Dartmouth College

Daniel L. Solomon
North Carolina State University

H. Jean Thiebaux
National Science Foundation

Ronald A. Thisted
University of Chicago

Gracia Toubia
Wichita, Kansas

John R. Tucker
National Research Council

Richard Tweedie
Colorado State University

Joe H. Ward
University of Texas at San Antonio